Developing Numeracy
MEASURES, SHAPE AND SPACE
ACTIVITIES FOR THE DAILY MATHS LESSON

year

Hilary Koll and Steve Mills

A & C BLACK

Contents

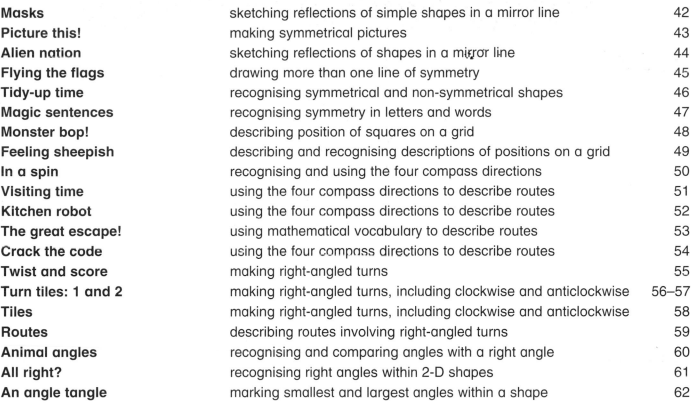

Published 2001 by A & C Black (Publishers) Limited
37 Soho Square, London W1D 3QZ

ISBN 0-7136-5878-9

Copyright text © Hilary Koll and Steve Mills, 2001
Copyright illustrations © Michael Evans and Leon Baxter, 2001
Copyright cover illustration © Charlotte Hard, 2001
Editors: Lynne Williamson and Marie Lister

The authors and publishers would like to thank Madeleine Madden and Corinne McCrum for their advice in producing this series of books.

A CIP catalogue record for this book is available from the British Library.

Printed in Great Britain by Caligraving Ltd, Thetford, Norfolk.

Introduction

Developing Numeracy: Measures, Shape and Space is a series of seven photocopiable activity books designed to be used during the daily maths lesson. They focus on the fourth strand of the National Numeracy Strategy *Framework for teaching mathematics*. The activities are intended to be used in the time allocated to pupil activities; they aim to reinforce the knowledge, understanding and skills taught during the main part of the lesson and to provide practice and consolidation of the objectives contained in the framework document.

Year 3 supports the teaching of mathematics by providing a series of activities which develop essential skills in measuring and exploring pattern, shape and space. On the whole the activities are designed for children to work on independently, although this is not always possible and occasionally some children may need support.

Year 3 encourages children to:

- read and begin to write the language of measure, shape, space and time;
- estimate, measure and compare length, mass, capacity and time, and to suggest suitable units and equipment for such measurements;
- read a scale to the nearest division, and to draw and measure lines to the nearest half centimetre;
- begin to use decimal notation for metres and centimetres;
- classify and describe 3-D and 2-D shapes referring to their properties;
- identify and draw lines of symmetry in simple shapes;
- read and begin to write the vocabulary related to position, direction and movement, and use the four compass directions;
- identify right angles in 2-D shapes.

Extension

Many of the activity sheets end with a challenge (**Now try this!**) which reinforces and extends the children's learning, and provides the teacher with the opportunity for assessment. On occasion, you may wish to read out the instructions and explain the activity before the children begin working on it. The children may need to record their answers on a separate piece of paper.

Organisation

Very little equipment is needed, but it will be useful to have available rulers, scissors, coloured pencils, interlocking cubes, string, counters, solid shapes, dice, small mirrors and glue. You will need to provide tape measures for page 6, tinned grocery items for page 14, a selection of at least five containers for page 19, this year's calendar for page 25, and dice labelled A to F for page 49.

The children should also have access to measuring equipment to give them practical experience of length, mass and capacity.

To help teachers to select appropriate learning experiences for the children, the activities are grouped into sections within each book. However, the activities are not expected to be used in that order unless otherwise stated. The sheets are intended to support, rather than direct, the teacher's planning.

Some activities can be made easier or more challenging by masking and substituting some of the numbers. You may wish to re-use some pages by copying them onto card and laminating them, or by enlarging them onto A3 paper.

Teachers' notes

Very brief notes are provided at the foot of each page giving ideas and suggestions for maximising the effectiveness of the activity sheets. These can be masked before copying.

Structure of the daily maths lesson

The recommended structure of the daily maths lesson for Key Stage 2 is as follows:

Start to lesson, oral work, mental calculation	5–10 minutes
Main teaching and pupil activities *(the activities in the **Developing Numeracy** books are designed to be carried out in the time allocated to pupil activities)*	about 40 minutes
Plenary *(whole-class review and consolidation)*	about 10 minutes

Whole-class warm-up activities

The following activities provide some practical ideas which can be used to introduce or reinforce the main teaching part of the lesson.

Measures

Make a kilogram

Call out measurements and ask the children to tell you how many more grams are needed to make a kilogram, for example: *200 grams; how many more?* (Answer: 800 grams)

This could also be used for millilitres/litre, centimetres/metre, or metres/kilometre.

Reading a scale

Pin to the board a large picture of a jug with a scale marked in 100 ml intervals from 0–1000 ml and numbered 200, 400, 600, 800, 1000 ml. Make a small arrow and colour it red. Place the arrow in different positions on the scale and ask the children to identify the different measurements, for example 100 ml, 500 ml.

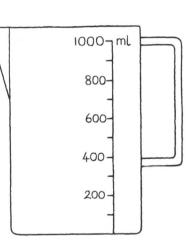

Call out further measurements and ask individuals to position the arrow correctly on the scale.

Estimating activities

Ask the children to estimate the length, mass or capacity of objects around the classroom, or use household groceries (with the actual measures masked over). Encourage them to suggest a range within which the measurement might fall, for example, between one and two metres. Once the class has agreed a range, the children can test their estimates by measuring.

Time cards

Discuss the different units of time and the relationships between them, for example, 1 week = 7 days, 1 day = 24 hours. Make a set of 12 cards showing the following times: 1 year, 365 days, 52 weeks, 12 months, 1 week, 7 days, 1 day, 24 hours, 1 hour, 60 minutes, 1 minute, 60 seconds. Hold up a pair of cards and ask the children to say which is longer or to say *Snap* if they show an equivalent time. More cards can be produced to extend the activity, for example, 15 minutes, 30 minutes, 45 minutes and the equivalent fraction cards.

Shape and space

2-D shape game

Draw a track on the board with sections large enough for children to draw shapes in. Draw a semi-circle in the first section on the track and ask the children to name the shape. Ask the children to work in pairs; one child names a flat shape and the other child must draw this shape in the next section of the track. Choose a new pair and continue.

Once all the spaces are filled in, describe the properties of one of the shapes and let the children identify it, for example: *It has two right angles, it has five sides, it is symmetrical*, and so on. Encourage the children to draw both regular and irregular shapes (those that do not have equal sides and equal angles).

I spy

Describe an object in the classroom, or which can be seen from the classroom window, and ask the children to guess the answer, for example: *I spy an object with six faces. All the faces are square. What is it?*

Twenty questions

Hide a shape in a bag and ask the children to find out which shape it is by asking questions. You can only answer *yes* or *no* to their questions. Challenge the children to guess the shape in twenty questions.

Noughts and crosses

Draw a 5 x 5 grid on the board and label the squares so that each can be identified with a letter and a number, for example, E3. Divide the class into two teams. One team crosses, the other circles. A child from each team should name a square in turn, for example, D4, and this should be marked with a cross or a circle. The aim is for one team to get four circles or crosses in a line.

Turning game

Ask the children to stand up and close their eyes. Call out instructions, such as: *Make a right angle turn* or *Make a quarter turn*. Encourage the children to turn slowly and not to look at others.

Vital statistics

- **Work with a friend.**
- **Measure these lengths on your own body.**

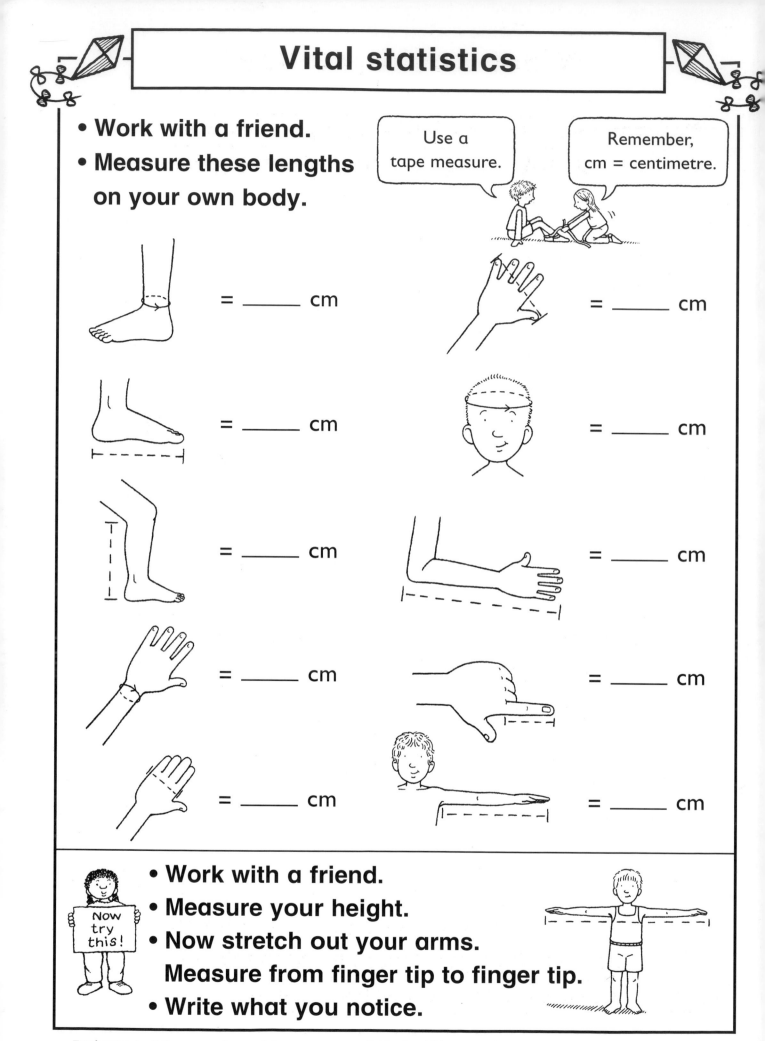

Use a tape measure.

Remember, cm = centimetre.

= _____ cm

= _____ cm

= _____ cm

= _____ cm

= _____ cm

= _____ cm

= _____ cm

= _____ cm

= _____ cm

= _____ cm

Now try this!

- **Work with a friend.**
- **Measure your height.**
- **Now stretch out your arms.**
  Measure from finger tip to finger tip.
- **Write what you notice.**

Teachers' note If there are not enough tape measures available, the children could be given a piece of string which they can mark and lay along a metre stick. Tell the children to give answers to the nearest centimetre. Encourage them to check that their measurements are reasonable, for example, 60 cm for the length of a finger must be wrong.

**Developing Numeracy
Measures, Shape and Space
Year 3**
© A & C Black 2001

What would you use?

- You can use ⟨kilometres⟩, ⟨metres⟩ or ⟨centimetres⟩ to measure these lengths. Write which is best.

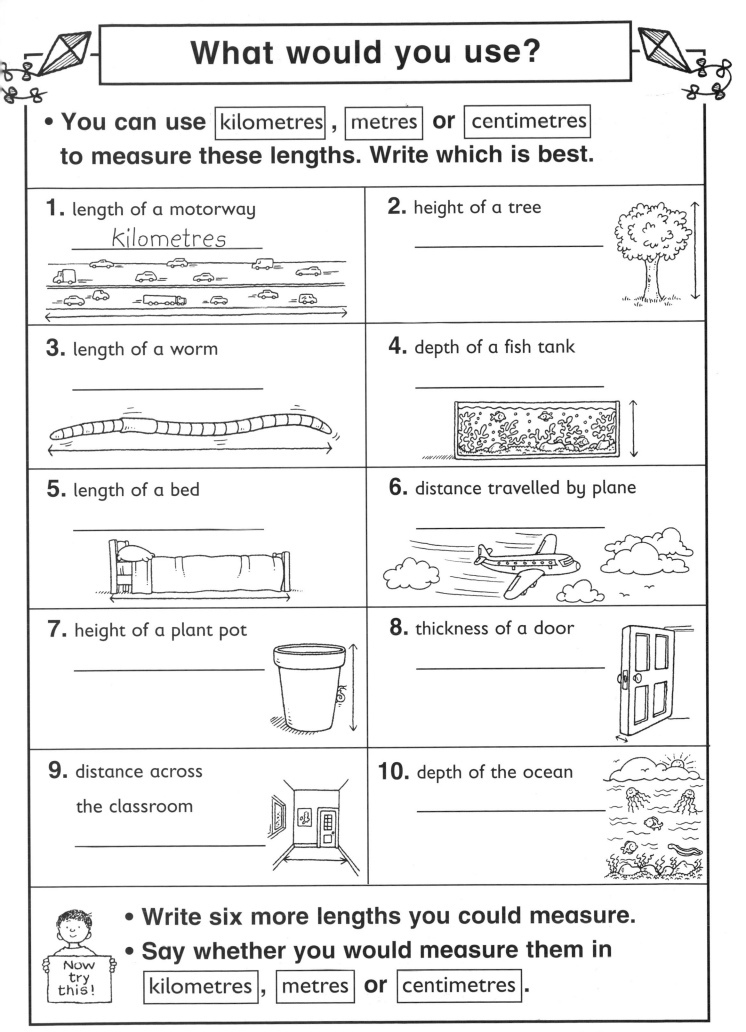

1. length of a motorway

kilometres

2. height of a tree

_____

3. length of a worm

_____

4. depth of a fish tank

_____

5. length of a bed

_____

6. distance travelled by plane

_____

7. height of a plant pot

_____

8. thickness of a door

_____

9. distance across the classroom

_____

10. depth of the ocean

_____

Now try this!

- **Write six more lengths you could measure.**
- **Say whether you would measure them in** ⟨kilometres⟩, ⟨metres⟩ or ⟨centimetres⟩.

Teachers' note The children can compare answers with a partner. Ensure that the children appreciate that the terms 'length', 'distance', 'height', 'width', 'depth' and 'thickness' are all lengths and can be measured using kilometres, metres and centimetres.

Developing Numeracy
Measures, Shape and Space
Year 3
© A & C Black 2001

Estimate extravaganza

- **Colour the best** estimate **to win the game show!**

Remember, m = metre.

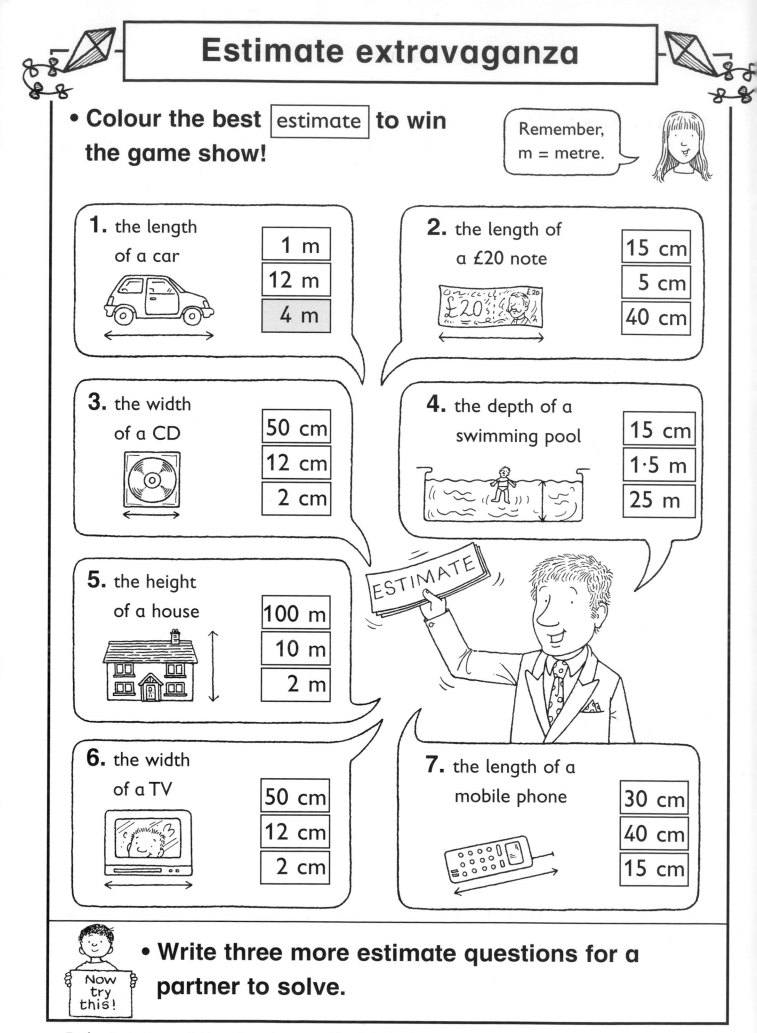

1. the length of a car
- 1 m
- 12 m
- 4 m

2. the length of a £20 note
- 15 cm
- 5 cm
- 40 cm

3. the width of a CD
- 50 cm
- 12 cm
- 2 cm

4. the depth of a swimming pool
- 15 cm
- 1·5 m
- 25 m

5. the height of a house
- 100 m
- 10 m
- 2 m

6. the width of a TV
- 50 cm
- 12 cm
- 2 cm

7. the length of a mobile phone
- 30 cm
- 40 cm
- 15 cm

Now try this!

- **Write three more estimate questions for a partner to solve.**

Teachers' note Have available a ruler and a metre stick to help the children visualise the lengths more clearly. Encourage the children to compare their answers with a partner and to discuss their reasons.

Developing Numeracy
Measures, Shape and Space
Year 3
© A & C Black 2001

Lengthy estimates

• **Circle the best** estimate **for each length.**

1. height of an elephant

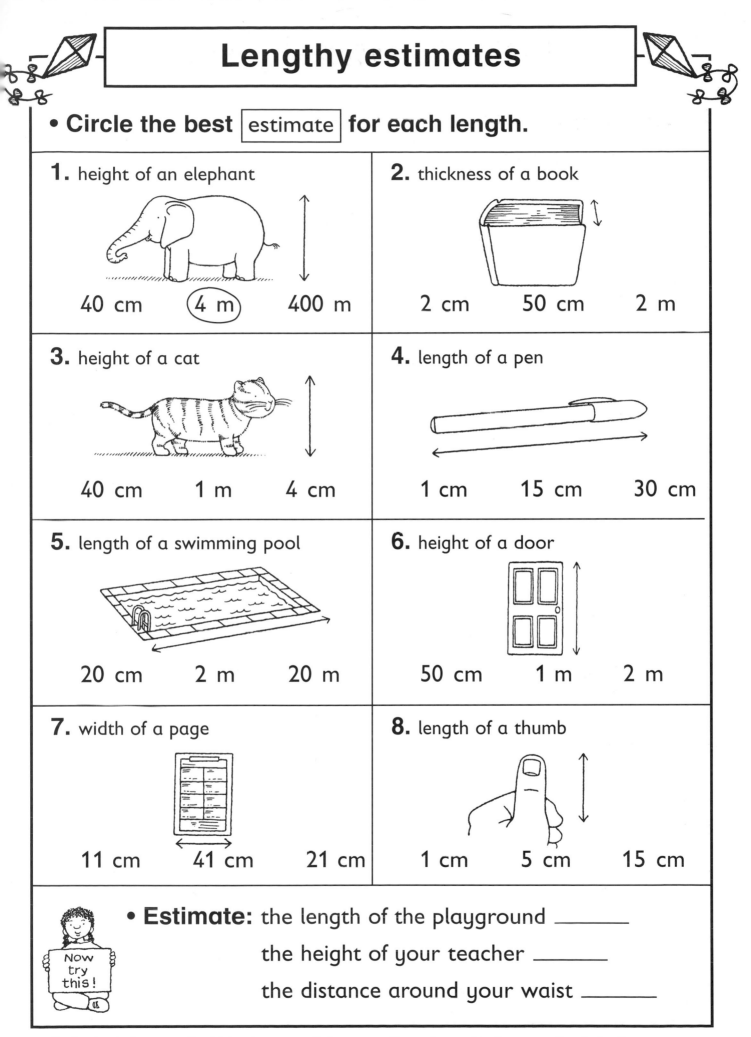

40 cm   (4 m)   400 m

2. thickness of a book

2 cm   50 cm   2 m

3. height of a cat

40 cm   1 m   4 cm

4. length of a pen

1 cm   15 cm   30 cm

5. length of a swimming pool

20 cm   2 m   20 m

6. height of a door

50 cm   1 m   2 m

7. width of a page

11 cm   41 cm   21 cm

8. length of a thumb

1 cm   5 cm   15 cm

• **Estimate:** the length of the playground _____

the height of your teacher _____

the distance around your waist _____

Now try this!

Teachers' note Encourage the children to compare their answers with a partner and to discuss their reasons. Where possible, the children could check their estimates by measuring. The extension activity provides a greater challenge, as the children are required to choose a suitable unit and decide upon the number of units for each length.

**Developing Numeracy
Measures, Shape and Space
Year 3
© A & C Black 2001**

Sports day

Each of these lengths can be written in two ways.

• Fill in the missing lengths.

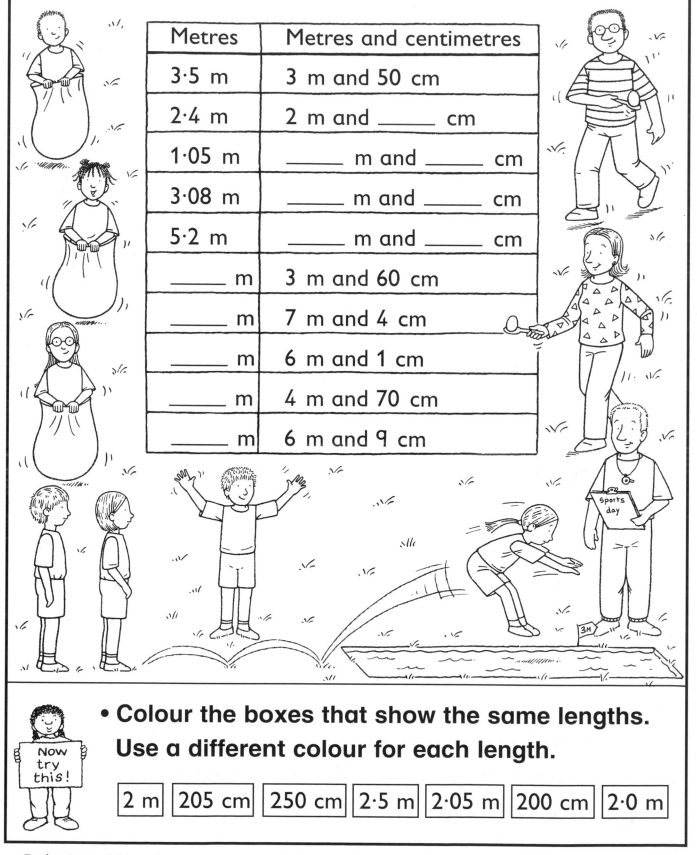

Metres	Metres and centimetres
3·5 m	3 m and 50 cm
2·4 m	2 m and _____ cm
1·05 m	_____ m and _____ cm
3·08 m	_____ m and _____ cm
5·2 m	_____ m and _____ cm
_____ m	3 m and 60 cm
_____ m	7 m and 4 cm
_____ m	6 m and 1 cm
_____ m	4 m and 70 cm
_____ m	6 m and 9 cm

• **Colour the boxes that show the same lengths.
Use a different colour for each length.**

| 2 m | 205 cm | 250 cm | 2·5 m | 2·05 m | 200 cm | 2·0 m |

Teachers' note Children often confuse measurements that are in decimal form, for example, 1·5 m and 1·05 m. When working with measurements with one decimal place, for example, 1·5 m, an extra zero can be added in the hundredths column (1·50 m), to help the children to see the number of centimetres more easily.

Developing Numeracy
Measures, Shape and Space
Year 3
© A & C Black 2001

Miniature garden

- **Use a ruler to measure the height of each plant.**
- **Give your answers in** centimetres .

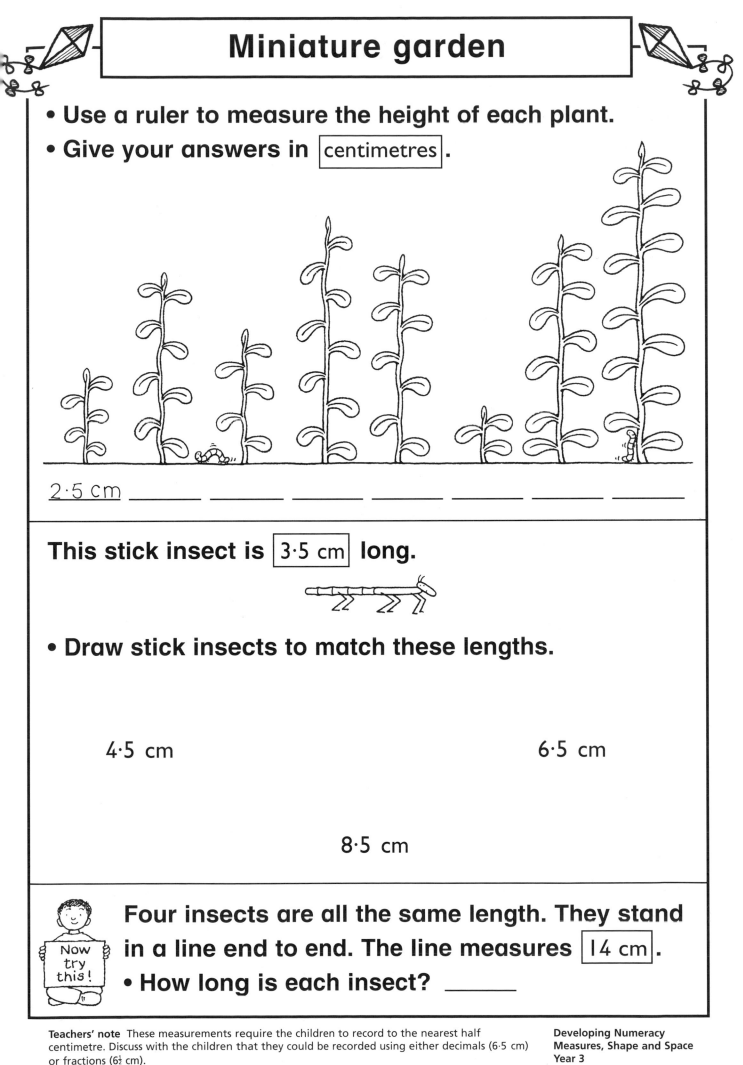

2·5 cm _____ _____ _____ _____ _____ _____ _____

This stick insect is 3·5 cm **long.**

- **Draw stick insects to match these lengths.**

4·5 cm   6·5 cm

8·5 cm

Four insects are all the same length. They stand in a line end to end. The line measures 14 cm .
- **How long is each insect?** _____

Teachers' note These measurements require the children to record to the nearest half centimetre. Discuss with the children that they could be recorded using either decimals (6·5 cm) or fractions (6½ cm).

Developing Numeracy
Measures, Shape and Space
Year 3
© A & C Black 2001

Lightning strikes

- **Measure the total length of each lightning strike in** centimetres **. Write the answer.**

Use a ruler.

1.

3 cm + 5 cm +
2 cm + 4 cm
= 14 cm

2.

3.

4.

5.

- **Draw your own lightning strike with a total length of** 30·5 cm **.**

Now try this!

Teachers' note The sections of the lightning strikes should be measured to the nearest centimetre or half centimetre. Encourage the children to record the latter using '·5', for example: 3·5 cm, 5·5 cm.

Developing Numeracy
Measures, Shape and Space
Year 3
© A & C Black 2001

Spaghetti scramble

- **With a partner, measure the length of each piece of spaghetti.**

Use a piece of string and a ruler.

1.

_____ cm

2.

_____ cm

3.

_____ cm

4.

_____ cm

5.

_____ cm

Now try this!

- **Draw a piece of spaghetti exactly 27 cm long.**

Teachers' note Demonstrate how to lay string along a curved line, mark it and straighten it along a ruler or metre stick to find the length of the line.

Developing Numeracy
Measures, Shape and Space
Year 3
© A & C Black 2001

Mass mix-up

- **What is the** mass **of each container?**
- **Colour the label which shows the best estimate.**

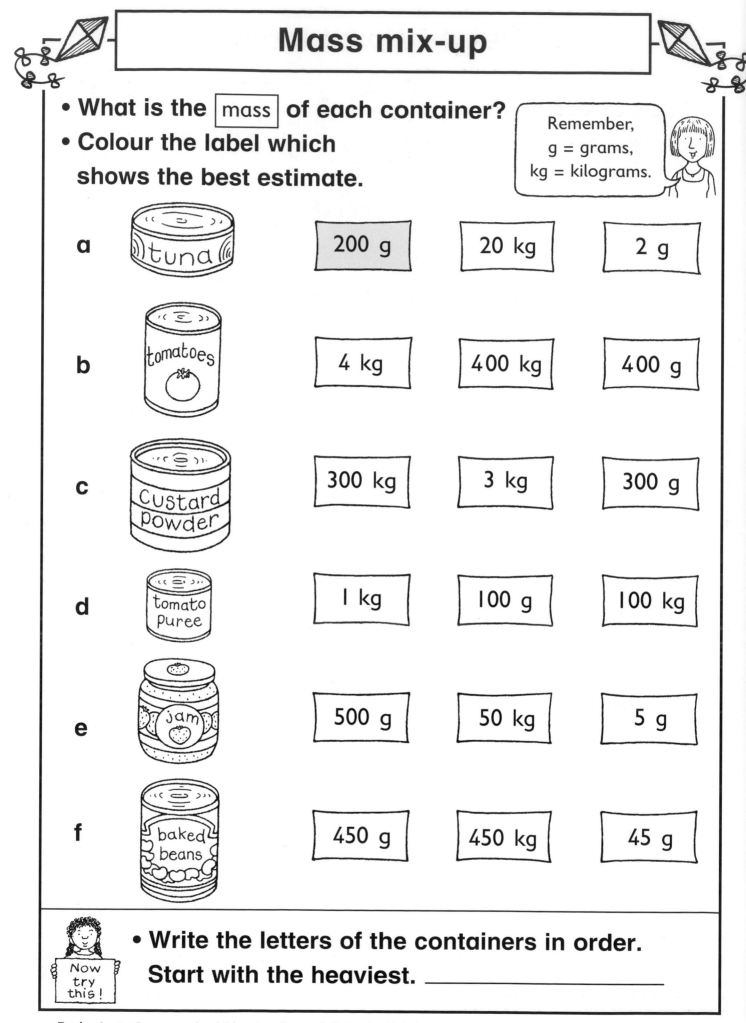

a tuna | 200 g | 20 kg | 2 g

b tomatoes | 4 kg | 400 kg | 400 g

c custard powder | 300 kg | 3 kg | 300 g

d tomato puree | 1 kg | 100 g | 100 kg

e jam | 500 g | 50 kg | 5 g

f baked beans | 450 g | 450 kg | 45 g

Now try this!

- **Write the letters of the containers in order. Start with the heaviest.** _____

Teachers' note Encourage the children to collect and display food labels to help them become aware of the masses of objects. Provide a collection of containers (such as those above) with the weights masked, for the children to hold and make estimates with.

Developing Numeracy
Measures, Shape and Space
Year 3
© A & C Black 2001

Weighty problems

- **Look at each set of weights.**
- **Write whether each set shows**

less than , more than **or** exactly **1 kilogram.**

Remember,
1000 grams = 1 kilogram.

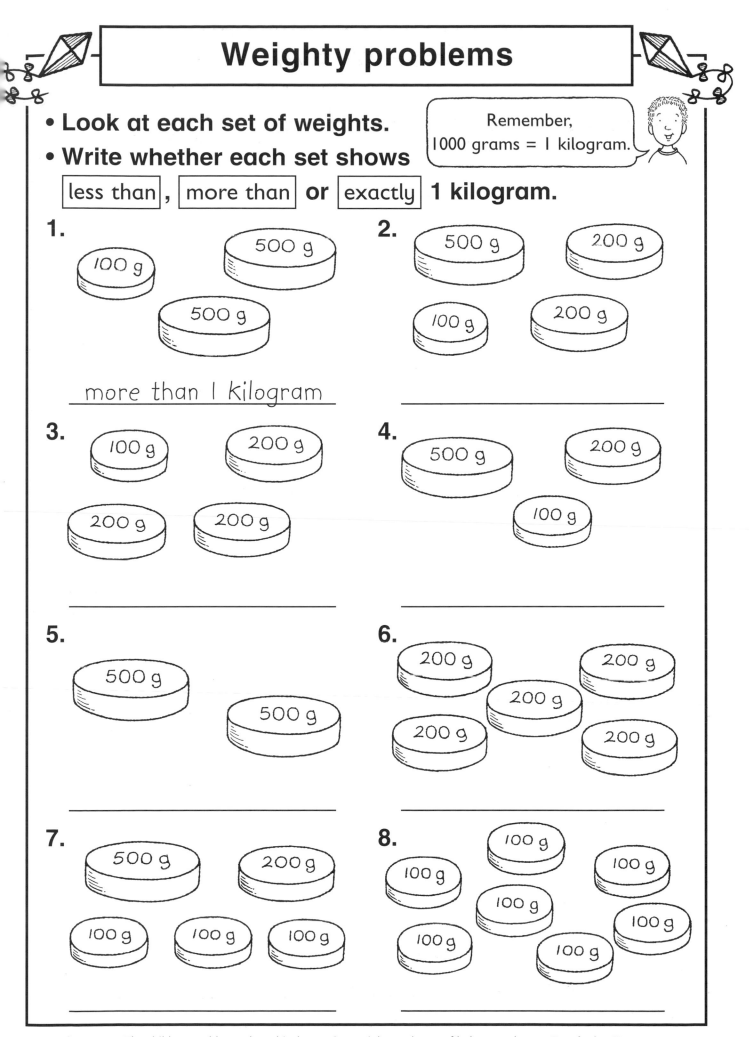

1.

100 g   500 g

500 g

more than 1 kilogram

2.

500 g   200 g

100 g   200 g

_____

3.

100 g   200 g

200 g   200 g

_____

4.

500 g   200 g

100 g

_____

5.

500 g

500 g

_____

6.

200 g   200 g

200 g

200 g   200 g

_____

7.

500 g   200 g

100 g   100 g   100 g

_____

8.

100 g

100 g   100 g

100 g

100 g   100 g

_____

Teachers' note The children could complete this sheet using weights and a set of balance scales, comparing the weights shown with a kilogram weight. As an extension, the children could investigate collections of weights which total 2 kg.

**Developing Numeracy
Measures, Shape and Space
Year 3**
© A & C Black 2001

Cooking for four

• **Read the recipes. Each recipe is for four people.**

Cheese scones
100 g flour
25 g butter
50 g cheese
300 ml milk

Cheese straws
50 g butter
150 g flour
100 g cheese
2 eggs

Chocolate meringues
30 g sugar
50 g chocolate
75 g cream
3 eggs

Crunchies
200 g flour
200 g butter
150 g sugar

Choc-chip cookies
250 g butter
150 g sugar
300 g flour
20 g chocolate

Chocolate muffins
$\frac{1}{2}$ kg flour
200 g sugar
500 ml milk
25 g chocolate
1 egg

Do not count recipes
that do not have
the ingredient.

1. Which recipe uses the **most**

flour? _____muffins_____   sugar? _____

milk? _____   butter? _____

chocolate? _____   cheese? _____

2. Which recipe uses the **least**

flour? _____   sugar? _____

milk? _____   butter? _____

chocolate? _____   cheese? _____

Teachers' note The children will need to use this page as a reference sheet when completing the activity on the following page. Remind the children that when they compare measurements they should look carefully at which unit is used, for example, 2 kg is greater than 200 g. Also explain to them that if a recipe does not have an ingredient it should not be included.

Developing Numeracy
Measures, Shape and Space
Year 3
© A & C Black 2001

Cooking for eight

- Use the recipes from 'Cooking for four'.
- Rewrite the measurements to make each recipe for eight people.

Cheese scones

_200 g_ flour

_____ butter

_____ cheese

_____ milk

Crunchies

_____ flour

_____ butter

_____ sugar

Cheese straws

_____ butter

_____ flour

_____ eggs

_____ cheese

Choc-chip cookies

_____ butter

_____ sugar

_____ flour

_____ chocolate

Chocolate meringues

_____ eggs

_____ sugar

_____ chocolate

_____ cream

Chocolate muffins

_____ flour

_____ sugar

_____ milk

_____ chocolate

_____ eggs

Now try this!

- Rewrite three recipes from 'Cooking for four'. Make each recipe for two people.

Teachers' note Revise doubling and halving strategies before starting this activity. The children will need to use the previous page as a reference sheet. Remind the children to include the units when writing measurements.

Developing Numeracy
Measures, Shape and Space
Year 3
© A & C Black 2001

Mix and match

- ● **Cut out the cards.**
- ● **Match pairs which show the same** | mass | .

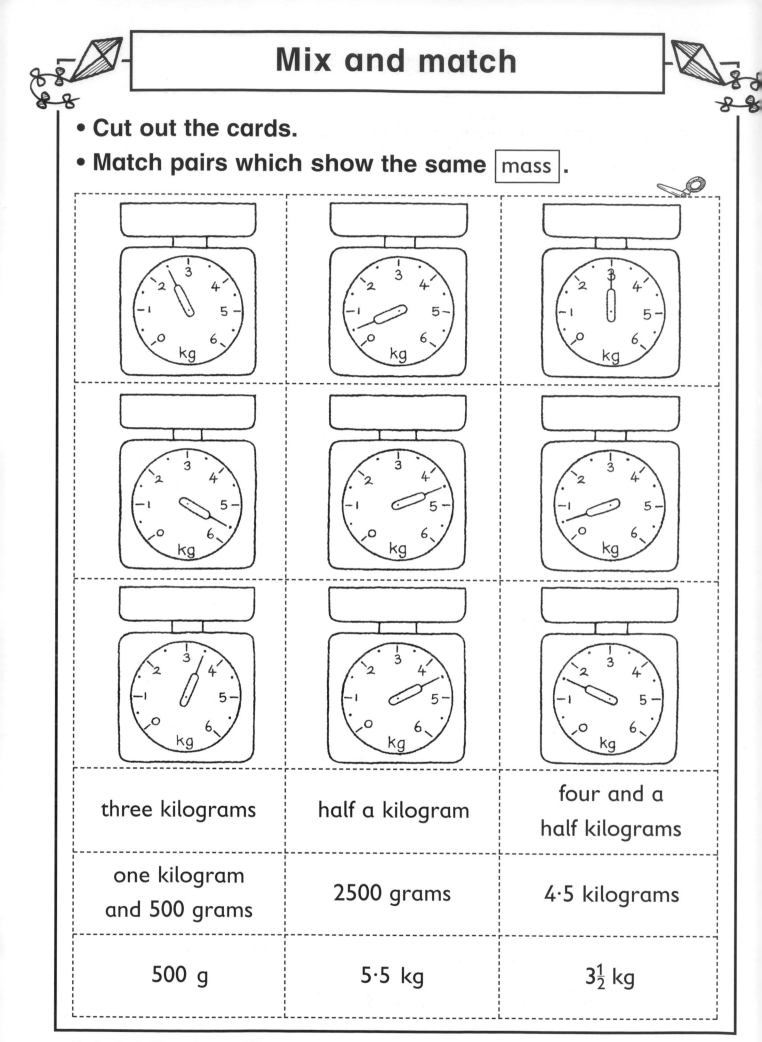

three kilograms	half a kilogram	four and a half kilograms
one kilogram and 500 grams	2500 grams	4·5 kilograms
500 g	5·5 kg	$3\frac{1}{2}$ kg

Teachers' note Discuss the many different ways to represent mass, using words, fractions, decimals, mixed units, and so on. The children could also use the cards to play Snap or Pelmanism.

Developing Numeracy
Measures, Shape and Space
Year 3
© A & C Black 2001

Waterslide

- **You can use** `millilitres` **or** `litres` **to measure the capacity of these containers. Write which is best.**

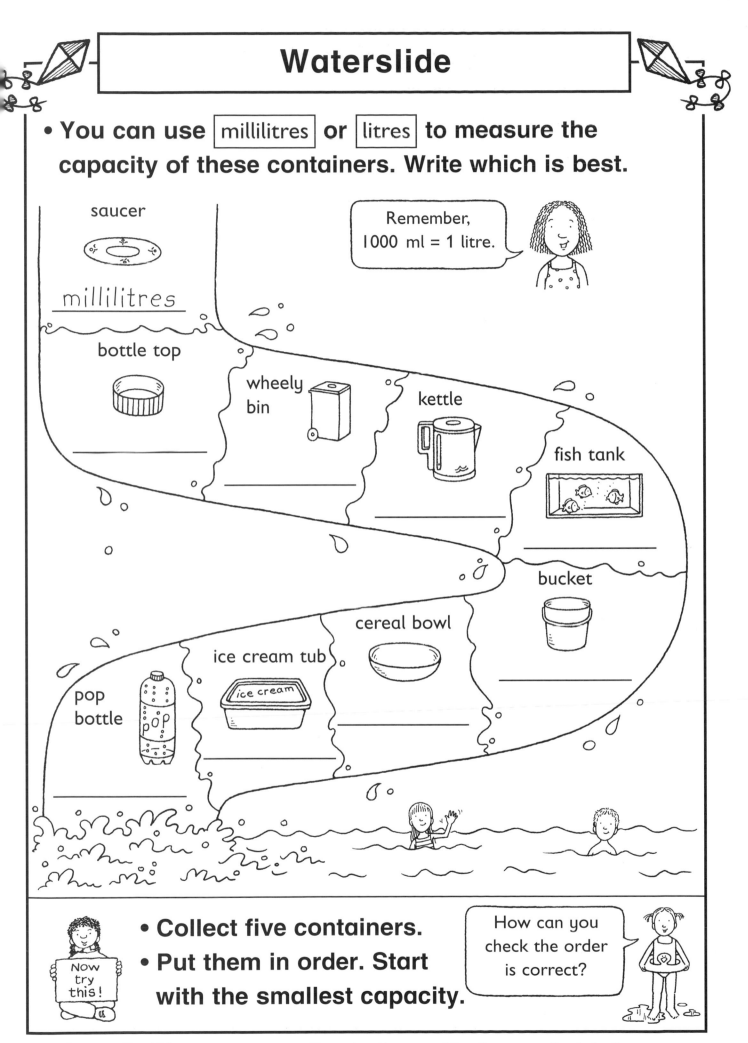

saucer

millilitres

Remember,
1000 ml = 1 litre.

bottle top

wheely bin

kettle

fish tank

bucket

cereal bowl

ice cream tub

pop bottle

- **Collect five containers.**
- **Put them in order. Start with the smallest capacity.**

Now try this!

How can you check the order is correct?

Teachers' note The children can compare answers with a partner. Have some of these items available for the children to test, using water or sand. Some children may need 10 ml, 100 ml and 1 litre containers for reference.

**Developing Numeracy
Measures, Shape and Space
Year 3**
© A & C Black 2001

What do you think?

The children are estimating the capacity of containers.

• Colour the estimate you think is best.

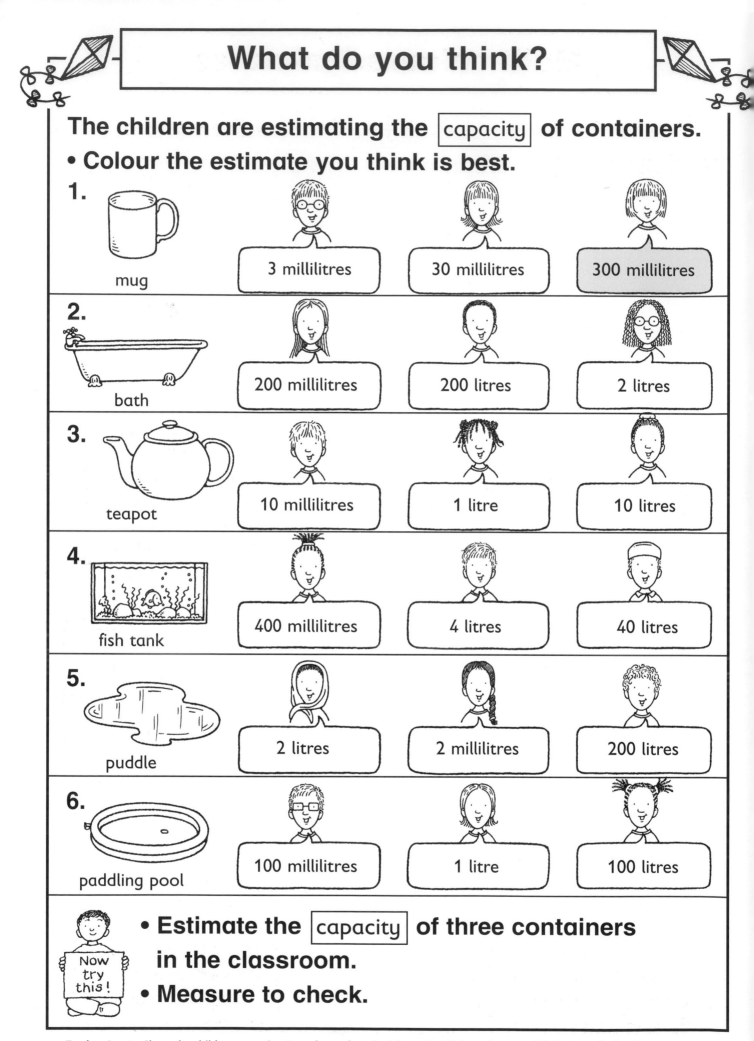

1. mug
- 3 millilitres
- 30 millilitres
- 300 millilitres

2. bath
- 200 millilitres
- 200 litres
- 2 litres

3. teapot
- 10 millilitres
- 1 litre
- 10 litres

4. fish tank
- 400 millilitres
- 4 litres
- 40 litres

5. puddle
- 2 litres
- 2 millilitres
- 200 litres

6. paddling pool
- 100 millilitres
- 1 litre
- 100 litres

Now try this!

• **Estimate the** capacity **of three containers in the classroom.**
• **Measure to check.**

Teachers' note Show the children a centimetre cube and remind them that if the cube were filled with water, this would be 1 millilitre. Then show them a 1000 Dienes block and help them to see that 1000 ml (or 1000 cm cubes) is the same as 1 litre. Some children may need 10 ml, 100 ml and 1 litre containers for reference.

Developing Numeracy
Measures, Shape and Space
Year 3
© A & C Black 2001

20

Multi-coloured mixtures

• Colour the layers in each jug. Start at the bottom.

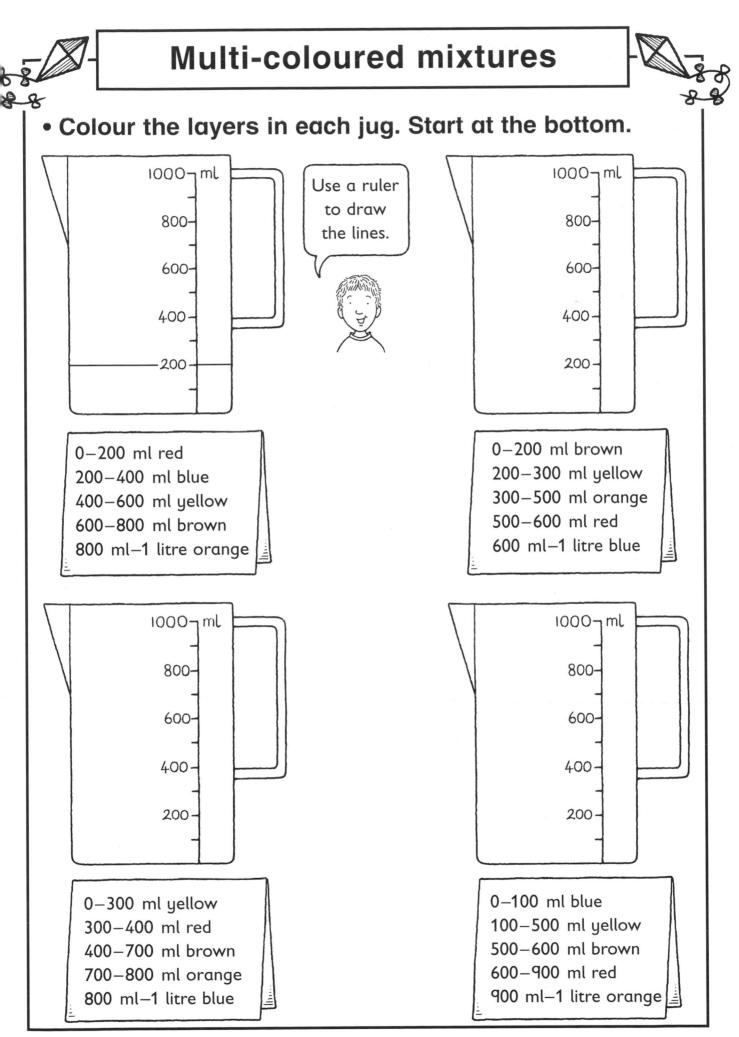

Use a ruler to draw the lines.

0–200 ml red
200–400 ml blue
400–600 ml yellow
600–800 ml brown
800 ml–1 litre orange

0–200 ml brown
200–300 ml yellow
300–500 ml orange
500–600 ml red
600 ml–1 litre blue

0–300 ml yellow
300–400 ml red
400–700 ml brown
700–800 ml orange
800 ml–1 litre blue

0–100 ml blue
100–500 ml yellow
500–600 ml brown
600–900 ml red
900 ml–1 litre orange

Teachers' note Encourage the children to use rulers and to draw horizontal lines to mark each layer. They will need the following coloured pencils for this activity: red, orange, yellow, brown, blue.

**Developing Numeracy
Measures, Shape and Space
Year 3
© A & C Black 2001**

21

Read all about it!

- **Join each newspaper headline to the correct unit.**

New long jump record – man jumps 11 _____

metres

Tiny baby weighs 900 _____

kilograms

millilitres

Lady in accident loses 900 _____ of blood

centimetres

grams

Broken pipe spills hundreds of _____ of water

litres

kilometres

metres

30 _____ of snow fall in Scotland

Man loses 20 _____ on diet

Motorway traffic jam 16 _____ long

Tallest man recorded at $2\frac{1}{2}$ _____

Now try this!

- **Write three newspaper headlines of your own. Use three different units.**

Teachers' note Spend some time discussing these units and categorising them into groups with the headings 'length', 'capacity' and 'mass'. Discuss that the words 'weight' and 'mass' are sometimes used to mean the same. Some children may need various types of measuring equipment to refer to.

Developing Numeracy
Measures, Shape and Space
Year 3
© A & C Black 2001

Space travellers

- **Join the matching times.**

1 year	7 days	60 seconds	1 hour
12 months	1 minute	1 week	24 hours
30 seconds	about 4 weeks	1 day	60 minutes
½ minute	1 century	1 month	52 weeks
		100 years	14 days
		1 year	2 weeks

- **Fill in the missing numbers.**

NOW try this!

2 hours = _____ minutes

2 days = _____ hours

2 years = _____ weeks

½ hour = _____ minutes

½ year = _____ months

Teachers' note Discuss these units of time with the children at the start of the lesson. Some children could work with dictionaries to check their answers.

Developing Numeracy
Measures, Shape and Space
Year 3
© A & C Black 2001

23

How long does it take?

- **Cut out the cards.**
- **Match each picture to the time it takes.**

wink	brush teeth	boil an egg
play a football match	sleep at night	watch a film
go on holiday	live a century	use up a calendar
4 minutes	2 weeks	1 second
100 years	2 minutes	2 hours
90 minutes	1 year	8 hours

Teachers' note The children could discuss their choices with a partner or a small group. As an extension activity, the children could put the cards in order. As a further extension, the children could make a timeline of their day, or of a person's life, to gain an appreciation of the passing of different units of time.

**Developing Numeracy
Measures, Shape and Space
Year 3**
© A & C Black 2001

24

This year

• **Use this year's calendar to answer the questions.**

1. What **day** of the week is:

3 October? _____   1 April? _____

25 December? _____   2 May? _____

19 March? _____   27 February? _____

20 June? _____   11 August? _____

18 January? _____   5 November? _____

21 July? _____   30 September? _____

2. What **date** is the:

first Tuesday in June? _____

second Sunday in October? _____

third Thursday in May? _____

last Friday in August? _____

• **In which** | months | **do these occur?**

Now try this!

Monday 6th _____

Saturday 22nd _____

Friday 24th _____

Teachers' note The children will need a copy of this year's calendar to answer these questions. They could tackle question 2 in either of two ways: dates could be written in words, for example, Tuesday, 5 June, 2001; or they could be written in numerical form: 05/06/01.

Developing Numeracy
Measures, Shape and Space
Year 3
© A & C Black 2001

Calendar month

Here is the month of April in the year 2003.

April 2003

M	T	W	T	F	S	S
	1	2	3	4	5	6
7	8	9	10	11	12	13
14	15	16	17	18	19	20
21	22	23	24	25	26	27
28	29	30				

1. Sam's birthday is on 4 April.

Which day of the week is this? _____

2. Sunita's birthday is 17/04/03.

On which day does her birthday fall? _____

3. Paul's birthday is on the first day of the month.

Write the date of his birthday in figures. _____

4. Write the dates of each Sunday in April.

06/04/03   _____   _____   _____

5. Write the date of the day before 01/04/03. _____

Which day of the week is this? _____

6. Write the date of the day after 30/04/03. _____

Which day of the week is this? _____

NOW try this! • **Write the calendar page for May 2003.**

Teachers' note Begin this lesson with revision of the months of the year. Ask questions such as: 'Which is the fourth month?' 'Which is the second month?' Revise different ways of writing dates, including the numerical form used above.

Developing Numeracy
Measures, Shape and Space
Year 3
© A & C Black 2001

Zoe's diary

- **Here are some pages from Zoe's diary.**
- **Write** |am| **or** |pm| **in the spaces.**

Monday

School trip this morning. We set off at 9:00 __am__ .

Tuesday

Was ill in the night at about 3:30 _____!

Wednesday

My birthday party is from 6:30 _____ to 8:00 _____.

Invitation 6·30 to 8·00

Thursday

Dentist this afternoon at 2:30 _____.

Yuk!

Friday

Going swimming today at 4:00 _____.

Saturday

Dad let me stay up until 10:30 _____ to watch a film.

Sunday

Slept at Gran's last night. Didn't get up until 9:30 _____.

Monday

Playtime at my new school is at 10:45 _____.

Tuesday

Sports day started at 2:00 _____.

- **Write what you are usually doing at:**

NOW try this!

3:00 am _____   3:00 pm _____

7:00 am _____   7:00 pm _____

9:00 am _____   9:00 pm _____

Teachers' note At the start of the lesson, draw a timeline showing the hours from midnight to midday (am) and from midday to midnight (pm). Encourage the children to discuss the types of activities they might be doing at various times throughout the day.

Developing Numeracy
Measures, Shape and Space
Year 3
© A & C Black 2001

27

Tell the time

• **Write these times in** words .

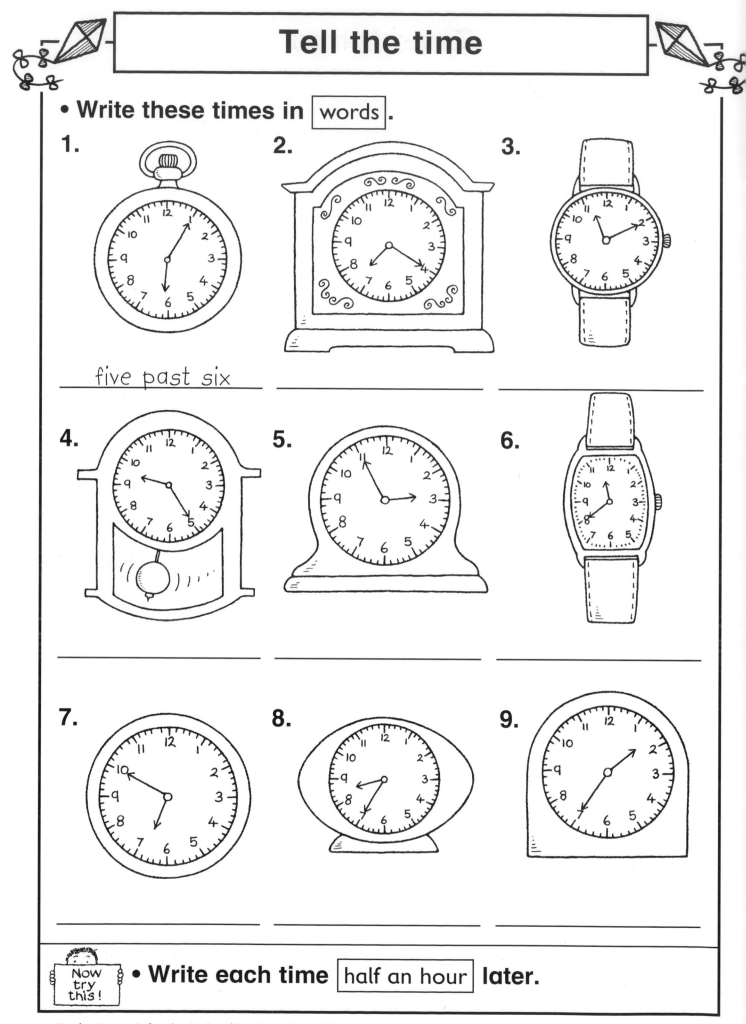

1.

_____
five past six

2.

_____

3.

_____

4.

_____

5.

_____

6.

_____

7.

_____

8.

_____

9.

_____

Now try this!

• **Write each time** half an hour **later.**

Teachers' note Before beginning this activity, the children should practise counting in fives and tens. The hands on these clocks can be masked to provide a flexible resource. Encourage the children to discuss what they might be doing at these times and to consider a range of activities from their own experiences, for example, the time school starts or ends.

Developing Numeracy
Measures, Shape and Space
Year 3
© A & C Black 2001

Time bingo

☆ Cut out the cards at the bottom of the page. Put them in a pile face down.

☆ Take turns to pick a card from the top of the pile.

☆ Cross off that time on your grid. Keep the card.

☆ If you cannot cross off the time, put the card to the bottom of the pile.

☆ The winner is the first to cross off all his or her times.

Play this game with a partner.

_____ 's grid

11:25	8:30	9:20
7:10	5:55	9:40
2:50	11:50	12:05

_____ 's grid

11:50	7:10	9:40
12:05	5:55	4:15
2:50	8:35	8:30

ten minutes to three	twenty past nine	half past eight	ten minutes past seven
quarter past four	five past twelve	twenty minutes to ten	ten past seven
five minutes to six	five minutes past twelve	ten to three	ten to twelve
twenty-five past eleven	thirty minutes past eight	twenty minutes past nine	five to six
ten minutes to twelve	twenty to ten	twenty-five to nine	ten past seven

Teachers' note The times can be masked and altered to provide a more flexible resource. Discuss the different ways of saying the same time, for example: ten to eight, ten minutes to eight, seven fifty.

Developing Numeracy Measures, Shape and Space Year 3
© A & C Black 2001

Clock cards

Developing Numeracy
Measures, Shape and Space
Year 3
© A & C Black 2001

☆ Cut out the clock cards.

☆ Spread them out face down.

☆ Take turns to reveal a card. Say the time.

☆ When you have turned over all the cards,
  sort them into pairs which show the same time.

Work with
a partner.

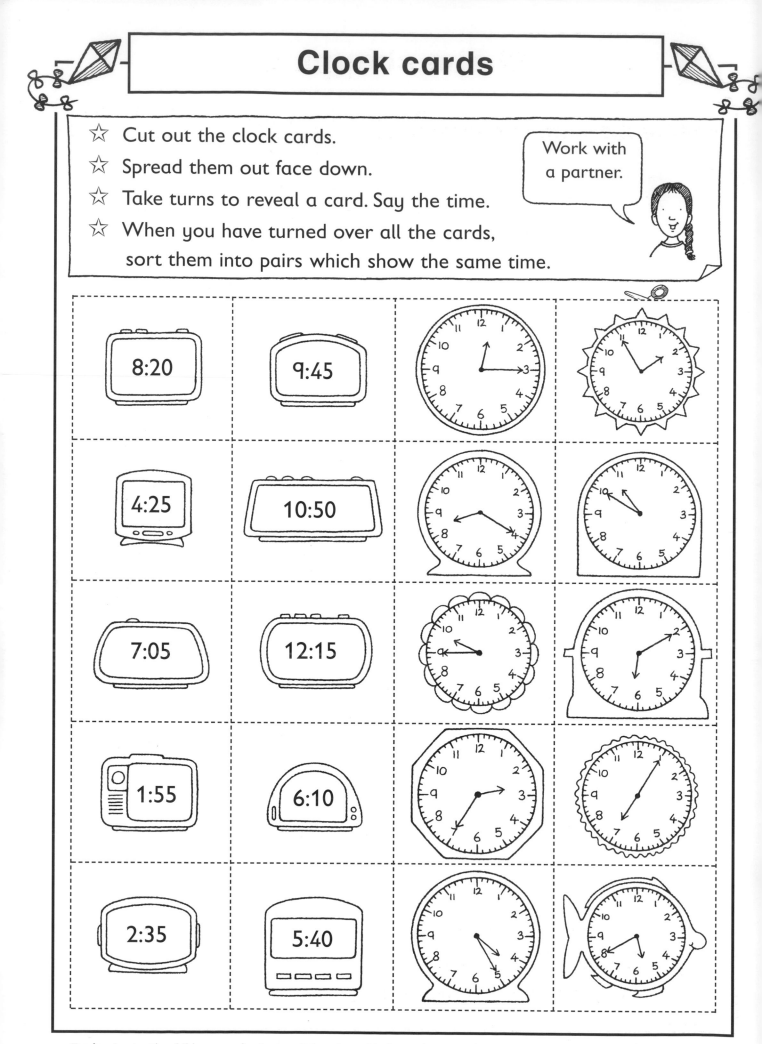

8:20

9:45

4:25

10:50

7:05

12:15

1:55

6:10

2:35

5:40

Teachers' note The children can play Snap or Pelmanism with the cards, or use them to practise telling the time with a partner. Alternatively, the children could work individually to match the pairs and stick the cards into their books, writing the times in words beneath each pair. They could also be asked to arrange either the analogue or digital cards in order.

Shape shopping

- **Write the name of each shape.**

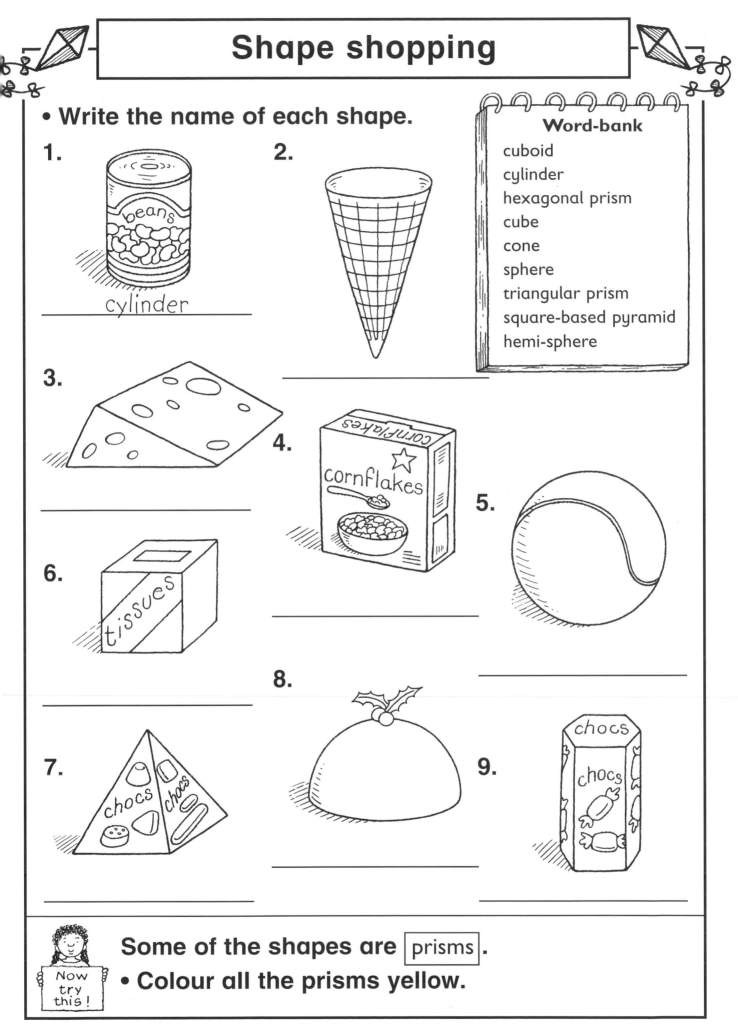

1. cylinder

2.

Word-bank
cuboid
cylinder
hexagonal prism
cube
cone
sphere
triangular prism
square-based pyramid
hemi-sphere

3.

4. cornflakes

5.

6. tissues

7. chocs chocs

8.

9. chocs chocs

Now try this!

Some of the shapes are prisms .
- **Colour all the prisms yellow.**

Teachers' note Revise 3-D shape names at the beginning of the lesson and introduce the term hemi-sphere.

Developing Numeracy
Measures, Shape and Space
Year 3
© A & C Black 2001

31

Shape descriptions

- **Work out which shapes the children are describing.**
- **Write the name and draw it.**

1.

My shape has two identical triangular faces. The other faces are rectangles.

triangular prism

2.

My shape has two circular faces and one curved face. It has two edges and no vertices.

3.

My shape has one vertex, one circular face and one curved face.

4.

My shape has a square base. The rest of the faces are triangles.

5.

My shape has six rectangular faces. Two of the faces are squares.

6.

My shape has one flat face and one curved face. It is half a sphere.

Now try this!

- **Hide a solid shape. Describe it to a partner.**
- **Can your partner guess the shape?**

Teachers' note Revise 'vertex', 'vertices' and 'hemi-sphere'. As some children may find drawing shapes very difficult, provide them with real shapes to help them draw the shapes on the page. Also provide some examples of 3-D shapes drawn on paper.

Developing Numeracy
Measures, Shape and Space
Year 3
© A & C Black 2001

Sorting solid shapes

- **Choose five solid shapes. Sort them onto the first Carroll diagram. Write the shape names under the headings. Repeat for the second Carroll diagram.**

Fewer than five faces	**Five or more faces**

At least one square face	**No square faces**

- **Draw a Carroll diagram with the headings** prisms **and** not prisms **. Add some shape names.**

Teachers' note Provide the children with a range of solid shapes and a word-bank with the names of the shapes. Carroll diagrams were devised by Lewis Carroll, who wrote *Alice in Wonderland.* It is important that the headings on each diagram are mutually exclusive, i.e. that all the shapes will fit under only one of the headings, for example, 'Red' or 'Not red'.

**Developing Numeracy
Measures, Shape and Space
Year 3**
© A & C Black 2001

What's the same?

- **Write two things that are the same about each pair of shapes.**

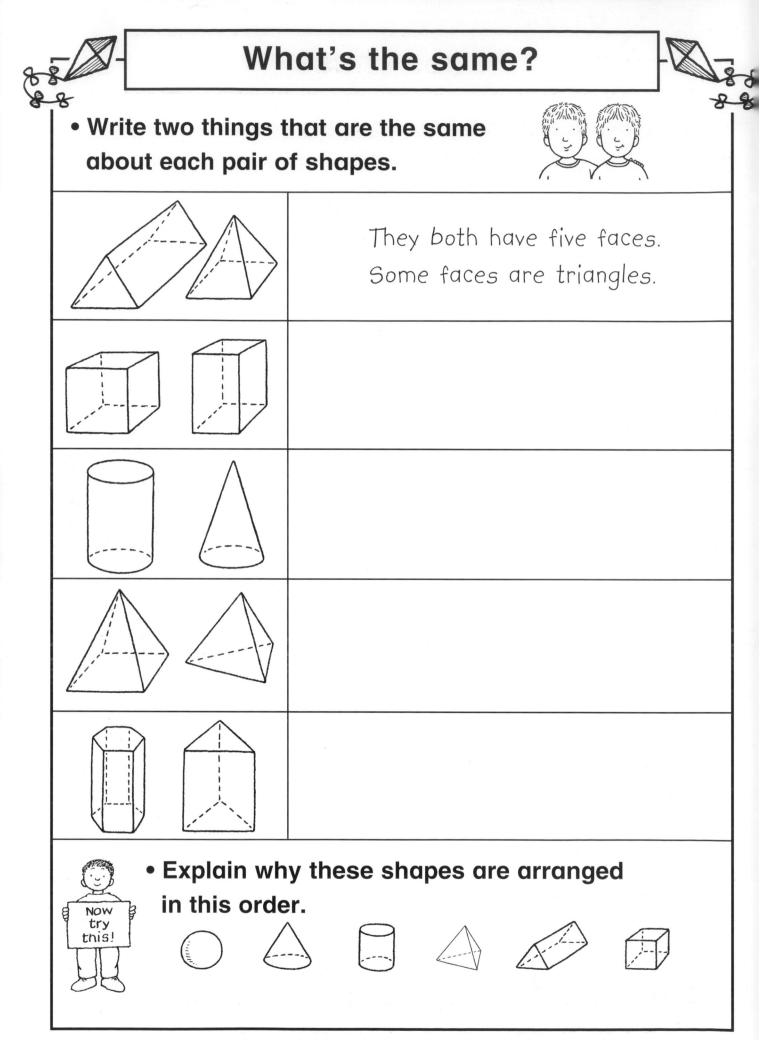

They both have five faces.
Some faces are triangles.

- **Explain why these shapes are arranged in this order.**

NOW try this!

Teachers' note Encourage the children to look at the number of faces, vertices and edges that each shape has, and the shapes of the faces. Other properties could also be discussed, including curved or flat faces and whether the faces have right angles.

Developing Numeracy
Measures, Shape and Space
Year 3
© A & C Black 2001

Cube chase

• **Read the instructions. Play the game with a partner.**

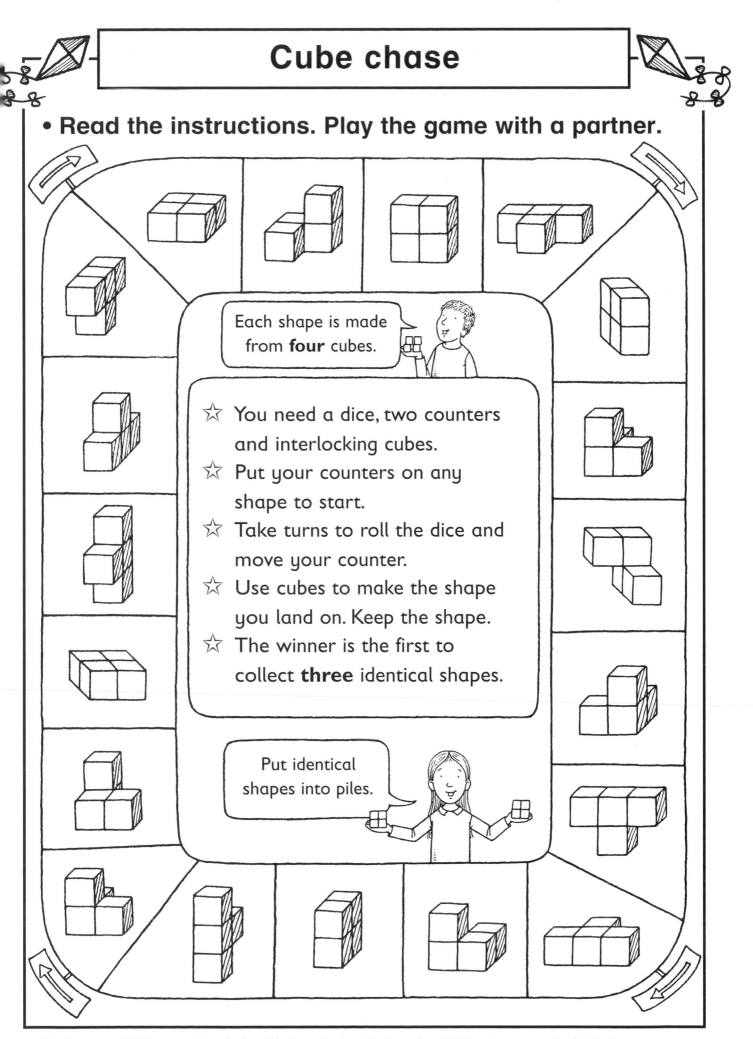

Each shape is made from **four** cubes.

☆ You need a dice, two counters and interlocking cubes.

☆ Put your counters on any shape to start.

☆ Take turns to roll the dice and move your counter.

☆ Use cubes to make the shape you land on. Keep the shape.

☆ The winner is the first to collect **three** identical shapes.

Put identical shapes into piles.

Teachers' note Children sometimes find it difficult to visualise 3-D shapes from 2-D drawings. Encourage them to hold the shape next to the drawing and to look at each cube in turn. As the children play the game, encourage them to sort identical shapes into piles.

Developing Numeracy Measures, Shape and Space Year 3 © A & C Black 2001

Crazy paving

• **Colour the crazy paving using the key.**

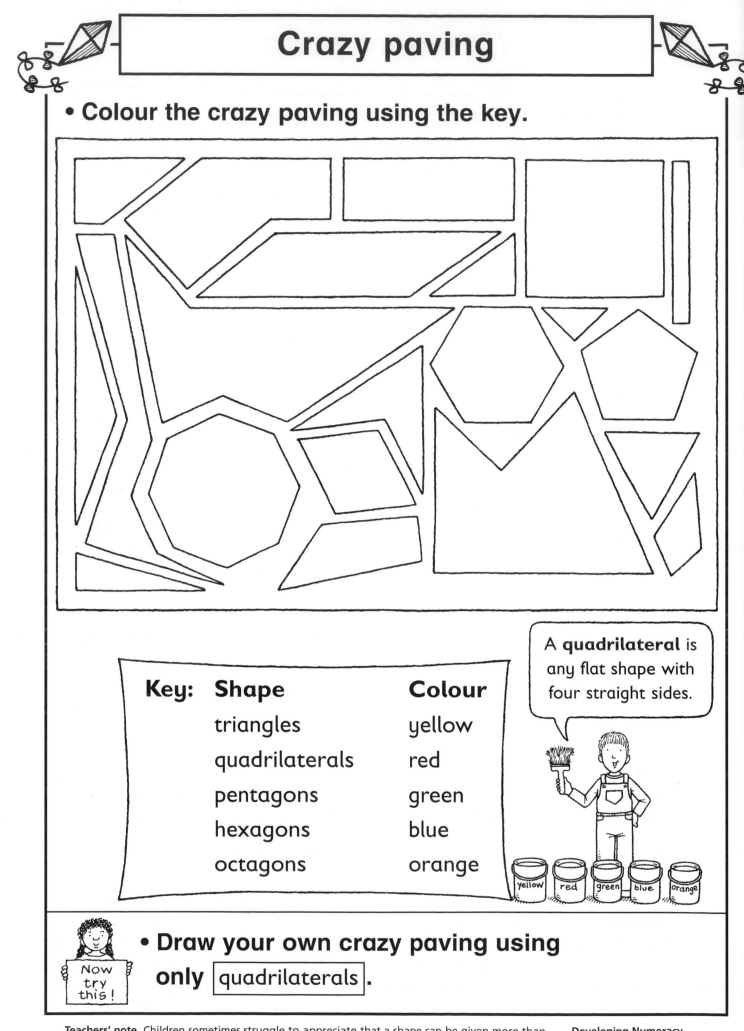

Key: **Shape**   **Colour**

Shape	Colour
triangles	yellow
quadrilaterals	red
pentagons	green
hexagons	blue
octagons	orange

A **quadrilateral** is any flat shape with four straight sides.

yellow   red   green   blue   orange

Now try this!

• **Draw your own crazy paving using only** quadrilaterals .

Teachers' note Children sometimes struggle to appreciate that a shape can be given more than one name, for example, that a square can be called a rectangle or a quadrilateral. Before the children begin the activity, discuss the different types of quadrilateral that they will come across. Remind the children of the difference between regular and irregular shapes.

Developing Numeracy
Measures, Shape and Space
Year 3
© A & C Black 2001

Matching pairs

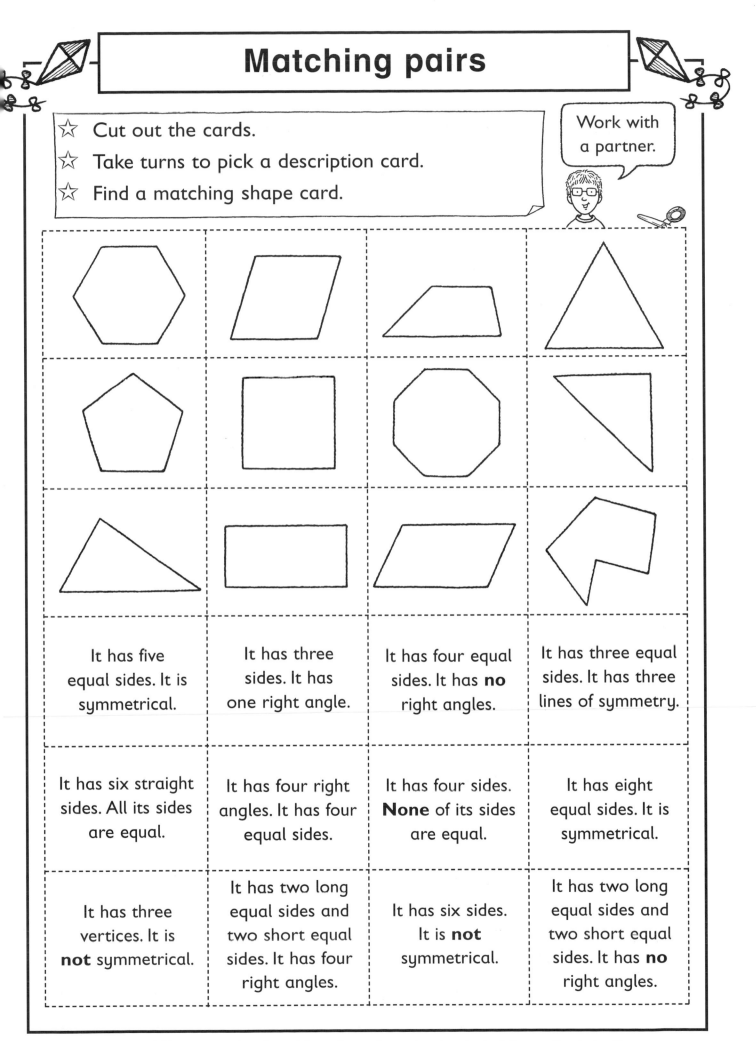

It has five equal sides. It is symmetrical.	It has three sides. It has one right angle.	It has four equal sides. It has **no** right angles.	It has three equal sides. It has three lines of symmetry.
It has six straight sides. All its sides are equal.	It has four right angles. It has four equal sides.	It has four sides. **None** of its sides are equal.	It has eight equal sides. It is symmetrical.
It has three vertices. It is **not** symmetrical.	It has two long equal sides and two short equal sides. It has four right angles.	It has six sides. It is **not** symmetrical.	It has two long equal sides and two short equal sides. It has **no** right angles.

Teachers' note The children may need a ruler, a right-angle template and a mirror to check the pairs. They could work individually to match the pairs and stick the cards in their books, or use the cards to play Snap or Pelmanism. Alternatively, the children could pick a shape card and describe it in detail, giving reference to sides, angles, vertices and symmetrical properties.

Developing Numeracy
Measures, Shape and Space
Year 3
© A & C Black 2001

Art gallery

• **Write two things about the shape in each painting.**

This triangle has one right angle.

It is not symmetrical.

• **Draw the shape.**

NOW try this!

This shape is symmetrical.

It has six sides. Two are longer than the rest.

Teachers' note Encourage the children to make reference to the number of sides and vertices, symmetrical properties, angles and lengths of sides.

Developing Numeracy Measures, Shape and Space Year 3 © A & C Black 2001

Stamp shapes

Here is a strip of stamps. It is a **rectangle** .

- **If two stamps are torn off, what shapes can the remaining stamps make?**

The remaining stamps must form one shape.

- **Shade the stamps you would tear off each time.**

octagon
_____   _____   _____

_____   _____   _____

NOW try this!

- **Write the shape remaining if two stamps are torn off these strips.**

_____   _____   _____   _____

Teachers' note Children sometimes miscount the sides of shapes like these. Provide them with a coloured pencil and ask them to draw a line along each side as they count it. They may need to be given the word 'decagon' for a 10-sided shape and, in the extension activity, 'dodecagon' for a 12-sided shape. The children can explore other arrangements of stamps in a similar way.

**Developing Numeracy
Measures, Shape and Space
Year 3**
© A & C Black 2001

Quad racing

☆ With a partner, take turns to roll a dice and move your counter.

You need: a dice, two counters and some cubes.

☆ If you land on a quadrilateral, collect a cube.

☆ The winner is the player who collects the most cubes.

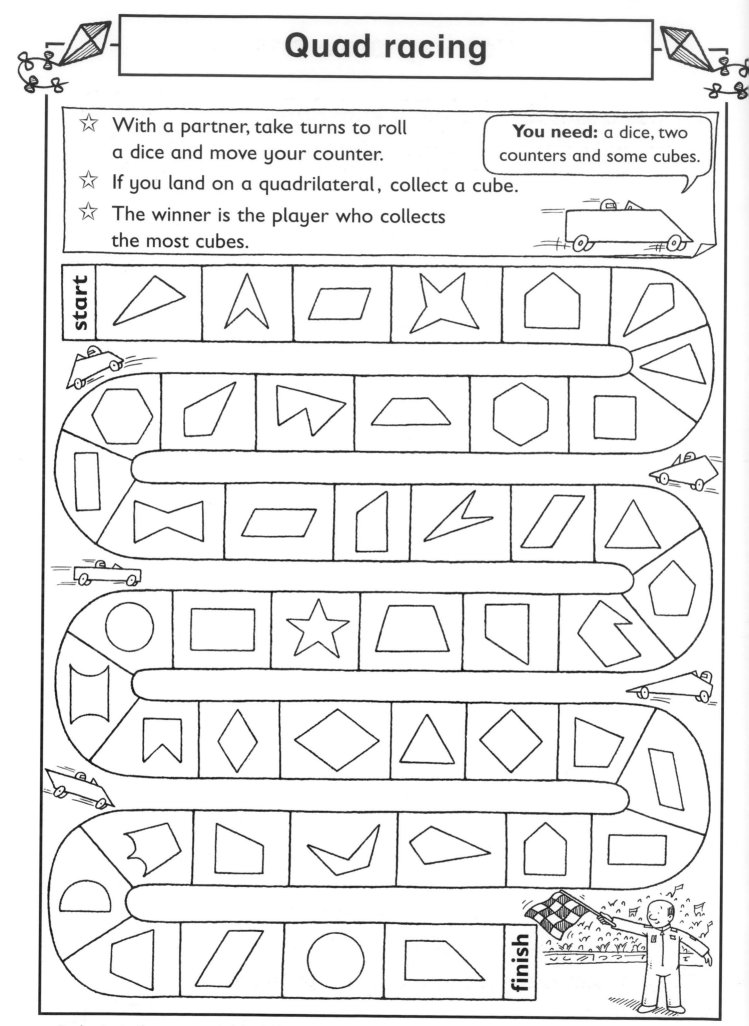

Teachers' note If necessary, remind the children that a quadrilateral is any flat shape with four straight sides. As an extension, the children could name the quadrilateral or describe another of its properties to gain an extra cube.

Developing Numeracy
Measures, Shape and Space
Year 3
© A & C Black 2001

Poultry patterns

These eggs have one line of symmetry .

- Complete and colour the pattern
  to make each egg symmetrical .

One has been done for you.

- This odd-shaped egg has two lines of symmetry .
- Complete and colour the pattern.

Now try this!

Teachers' note The children can use mirrors for this task, although some may find this difficult, failing to realise that the mirror must be lifted to see that the sketch underneath matches what can be seen in the mirror. Other perspex equipment is available that allows children to see the pattern through the perspex at the same time as seeing the reflection on the perspex.

Developing Numeracy
Measures, Shape and Space
Year 3
© A & C Black 2001

Masks

Ravi is making masks. He folds paper in half, cuts the shapes, then unfolds it.

One has been done for you.

- Look at the folded masks.
- Draw what each mask will look like when Ravi unfolds it.

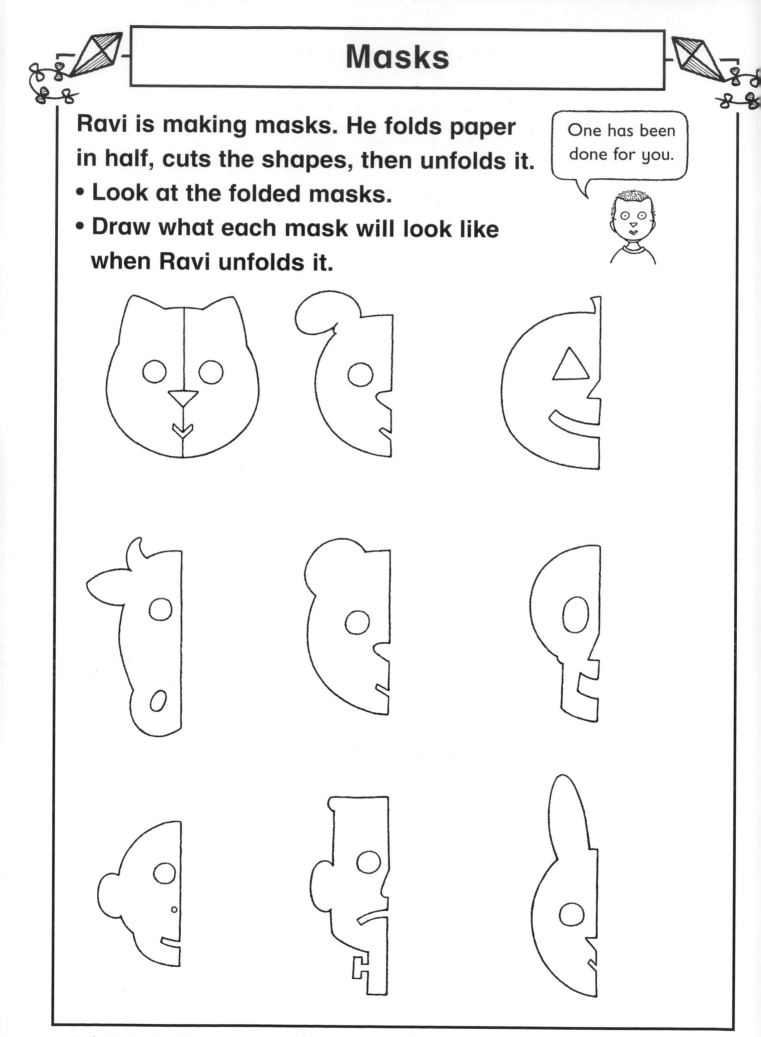

Teachers' note The children can use a mirror or tracing paper for this activity to check the reflections or to help them draw them more accurately. Emphasise, using examples, that certain points should be the same distance from the centre or the same length/width. As an extension, the children could make their own masks with one line of symmetry.

Developing Numeracy Measures, Shape and Space Year 3
© A & C Black 2001

Picture this!

This picture has | one line of symmetry |.

- **Cut out the hexagonal pieces and mix them up.**
- **Arrange the pieces to make the picture again.**

Teachers' note This sheet could be copied onto card and laminated to provide a more permanent resource. As an extension, the children could make their own symmetrical picture jigsaws on squared paper.

**Developing Numeracy
Measures, Shape and Space
Year 3**
© A & C Black 2001

43

Alien nation

On the planet Splog, all the aliens are symmetrical.

- Draw the reflection of each alien in the mirror line. Make it symmetrical.

One has been done for you.

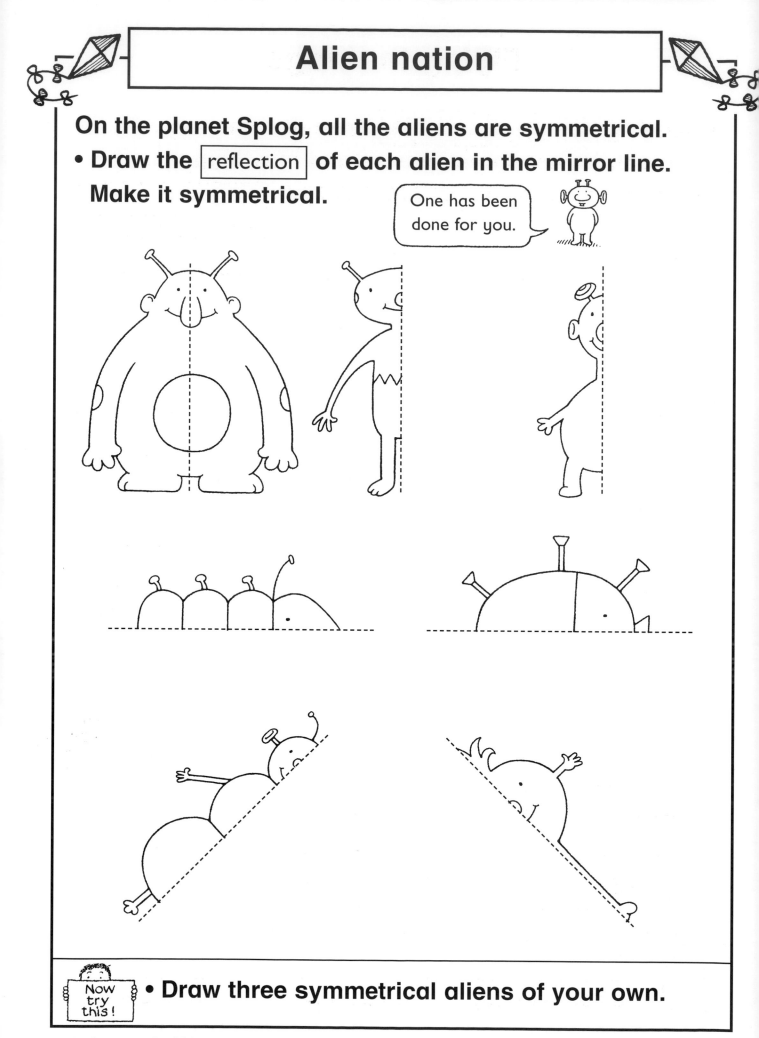

- Draw three symmetrical aliens of your own.

Teachers' note The children can use mirrors for this task, although some may find this difficult, failing to realise that the mirror must be lifted to see that the sketch underneath matches what can be seen in the mirror. Other perspex equipment is available that allows children to see the pattern through the perspex at reflection same time as seeing the reflection on the perspex.

**Developing Numeracy
Measures, Shape and Space
Year 3**
© A & C Black 2001

Flying the flags

Each flag has [0], [1] or [2] lines of symmetry.

• Draw the lines of symmetry. Write how many.

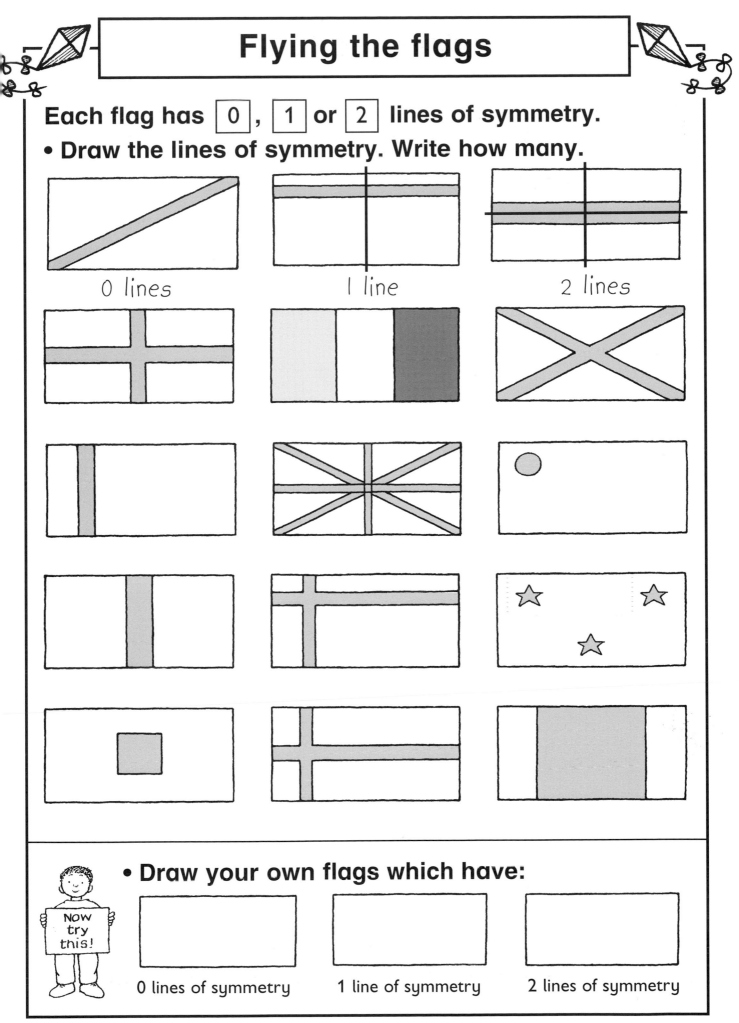

0 lines

1 line

2 lines

• Draw your own flags which have:

0 lines of symmetry

1 line of symmetry

2 lines of symmetry

Teachers' note Provide the children with small mirrors to help them check for lines of symmetry. Discuss and demonstrate that rectangles do not have a diagonal line of symmetry.

**Developing Numeracy
Measures, Shape and Space
Year 3**
© A & C Black 2001

Tidy-up time

The class has been cutting and folding paper shapes.

• Help to tidy up the shapes by joining each shape to the correct box.

• Use a ruler to draw lines of symmetry on the shapes.

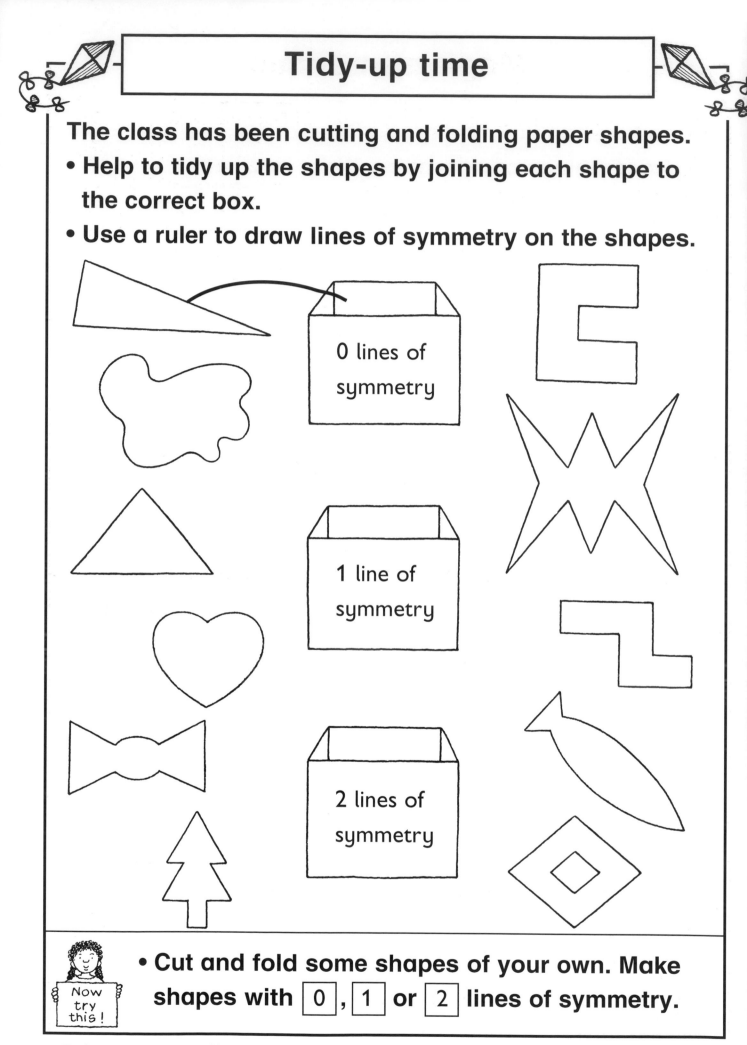

0 lines of symmetry

1 line of symmetry

2 lines of symmetry

• Cut and fold some shapes of your own. Make shapes with ⬚0⬚, ⬚1⬚ or ⬚2⬚ lines of symmetry.

Teachers' note This sheet could usefully follow a practical lesson where children have been cutting and folding paper shapes. Provide the children with small mirrors to help them check the lines of symmetry.

**Developing Numeracy
Measures, Shape and Space
Year 3**
© A & C Black 2001

Magic sentences

- **Use a mirror to find out what is special about these sentences.**

> Place the mirror along the dotted line.

```
T       A       W       A
O       H       H       T
M       O       A       A
        T       T       T
H               A       T
I       T               Y
T       O       A
        M       H       T
T       A       A       O
I       T       T       O
M       O               T
        O               H
```

They all use [symmetrical] **letters, like these.**

M T A I W H U Y O X V

- **Make new words using these letters. Write them** [vertically]. **Then place a mirror along the dotted lines.**

- **Make words using these letters.**

  B   E   C   H   I   D   O   X

- **Write them** [horizontally]. **Example:** H O O D

- **Place a mirror along the dotted lines.**

Now try this!

Developing Numeracy
Measures, Shape and Space
Year 3
© A & C Black 2001

Teachers' note The children will need small mirrors for this activity. Encourage them to investigate different words that can be made from these letters and to use them to form simple 'sentences'. The children could complete this activity in pairs.

Monster bop!

Here are some monster families at a monster bop.

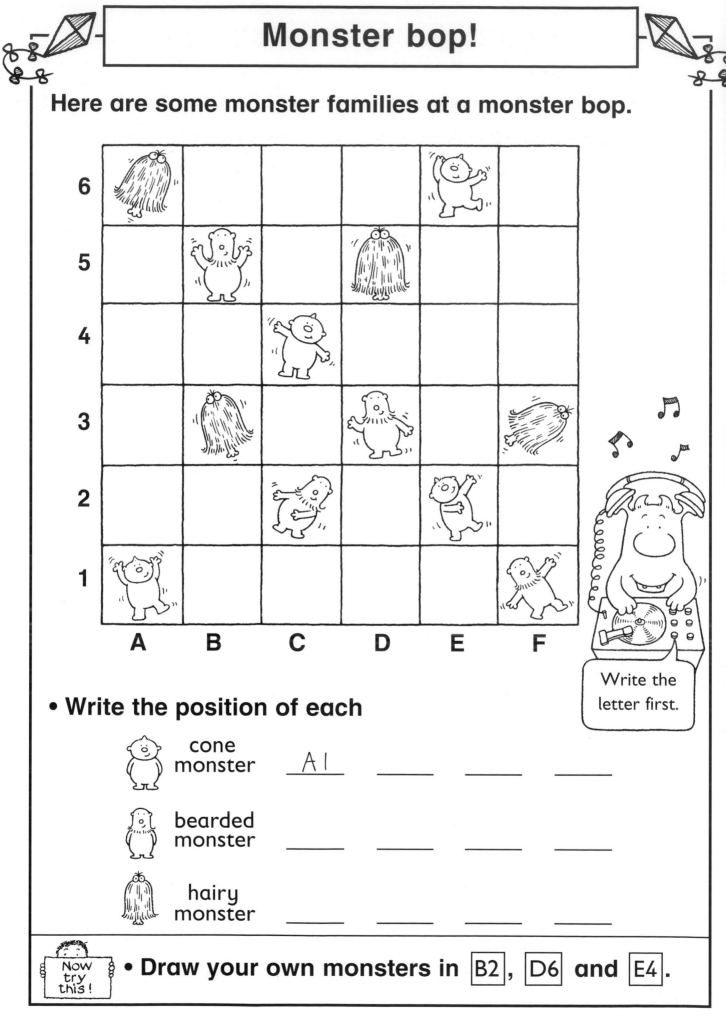

- **Write the position of each**

    cone
    monster   A1 ___ ___ ___ ___

    bearded
    monster   ___ ___ ___ ___

    hairy
    monster   ___ ___ ___ ___

Write the letter first.

- **Draw your own monsters in** B2 , D6 **and** E4 .

Now try this!

Teachers' note Remind the children that they should always write the letter before the number, for example, A5 rather than 5A.

Developing Numeracy
Measures, Shape and Space
Year 3
© A & C Black 2001

Feeling sheepish

• **Draw six sheep on the grid.**

> Put one sheep in a square.

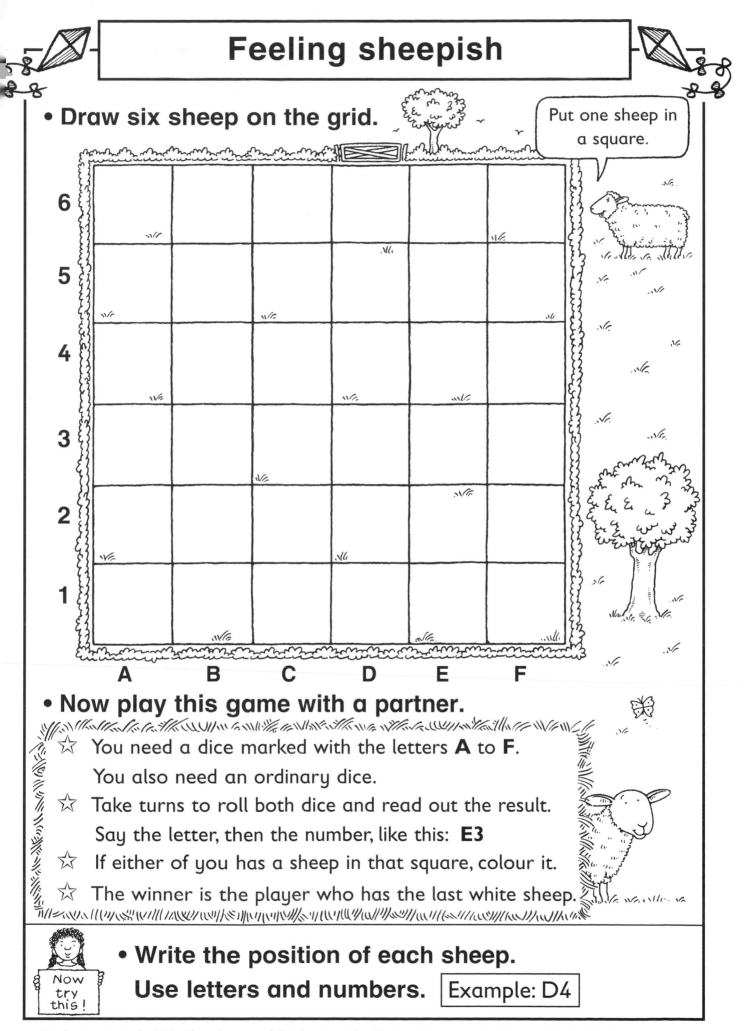

A   B   C   D   E   F

• **Now play this game with a partner.**

☆ You need a dice marked with the letters **A** to **F**.
   You also need an ordinary dice.

☆ Take turns to roll both dice and read out the result.
   Say the letter, then the number, like this: **E3**

☆ If either of you has a sheep in that square, colour it.

☆ The winner is the player who has the last white sheep.

Now try this!

• **Write the position of each sheep.**
Use letters and numbers. Example: D4

Teachers' note Each child will need a copy of this sheet and should play with a partner. Label a blank dice with the letters A–F, or stick small sticky labels over the faces of an ordinary dice.

Developing Numeracy
Measures, Shape and Space
Year 3
© A & C Black 2001

In a spin

- **First make the spinner at the bottom of the page.**

☆ Play this game with a partner. You need 13 counters and 2 cubes.

☆ Place the counters on the circles. Place the cubes on 'start'.

☆ Take turns to spin the spinner. Move your cube one square in that direction, if you can. If you land on a counter, collect it.

☆ The winner is the first player to collect five counters.

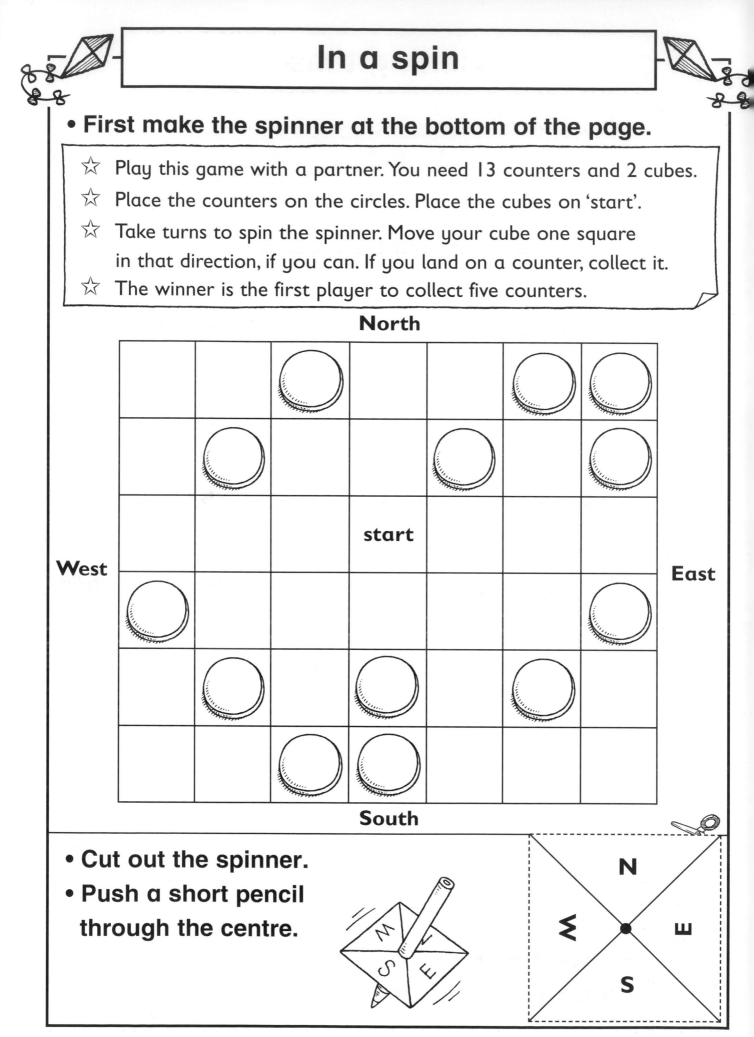

- **Cut out the spinner.**
- **Push a short pencil through the centre.**

Teachers' note Each pair of children needs one copy of the sheet. Discuss the compass directions with the children during the introduction. Ensure that the children understand the rules of the game and explain that if they cannot move in the direction shown, then they must miss a turn. They could make their spinners more durable by sticking them onto card (a biro top can be used instead of a pencil).

Developing Numeracy
Measures, Shape and Space
Year 3
© A & C Black 2001

50

Visiting time

This is a plan of the ground floor of a hospital.

Tom

Nurse

Eve

Alia

Jack

wall

Entrance ⇨

N
W — E
S

• **Starting at the entrance, write how to get:**

to Jack East 6 squares, North 1 square, West 1 square

from Jack to Alia _____

_____

from Alia to Tom _____

_____

from Tom to Eve _____

Now try this!

• **Write how the nurse can visit each patient in turn.**

Teachers' note Encourage the children to describe the routes using North, South, East and West. The more able children could also write instructions for the characters themselves, using phrases such as: face West; turn through one right angle anticlockwise; move forward two squares.

Developing Numeracy
Measures, Shape and Space
Year 3
© A & C Black 2001

Kitchen robot

This robot is programmed to go from the door to the freezer.

South 2, East 1, South 3, West 1

South 2 means South 2 squares.

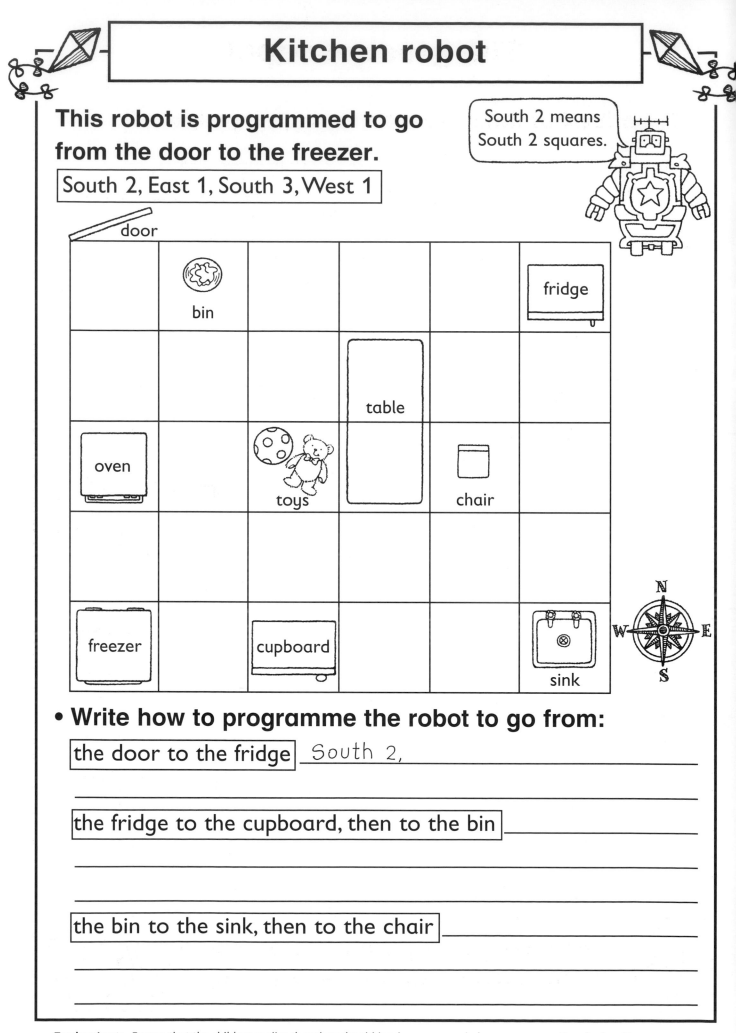

• **Write how to programme the robot to go from:**

the door to the fridge _South 2,_ _____

_____

the fridge to the cupboard, then to the bin _____

_____

_____

the bin to the sink, then to the chair _____

_____

_____

Teachers' note Ensure that the children realise that they should land on an occupied square when it is the robot's final destination, but should not cross occupied squares on the way. As an extension, ask the children to write instructions for a route where the robot visits every place in the kitchen once, starting at the door and finishing at the chair.

Developing Numeracy
Measures, Shape and Space
Year 3
© A & C Black 2001

The great escape!

• **Describe Pat the prisoner's escape from his cell.**

Use these words.

Word-bank

go	turn	forwards	backwards
ascend	descend	horizontally	vertically
clockwise	anticlockwise	left	right

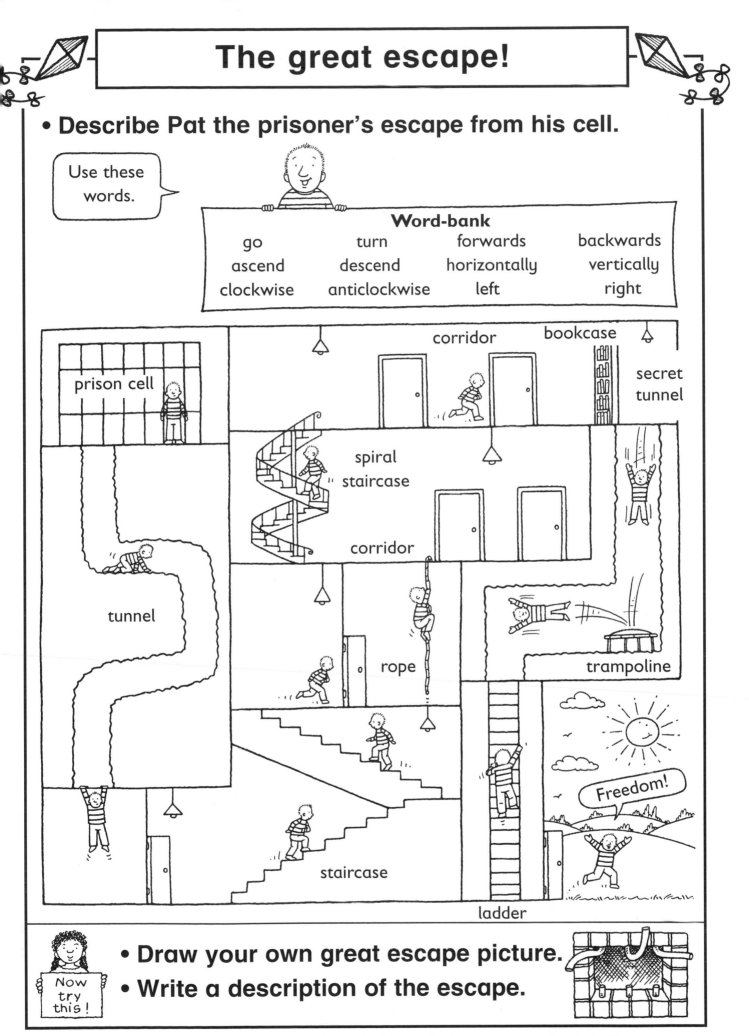

prison cell

corridor   bookcase

secret tunnel

spiral staircase

corridor

tunnel

rope

trampoline

staircase

ladder

Freedom!

Now try this!

• **Draw your own great escape picture.**
• **Write a description of the escape.**

Teachers' note Before the children begin the activity, discuss what is happening in the illustration. The children can describe the escape to a partner or they can record it on paper.

Developing Numeracy
Measures, Shape and Space
Year 3
© A & C Black 2001

Crack the code

Follow the instructions to reveal a book title.

N = North, S = South, E = East, W = West

Jump E3 means lift your pencil and jump East 3 squares.

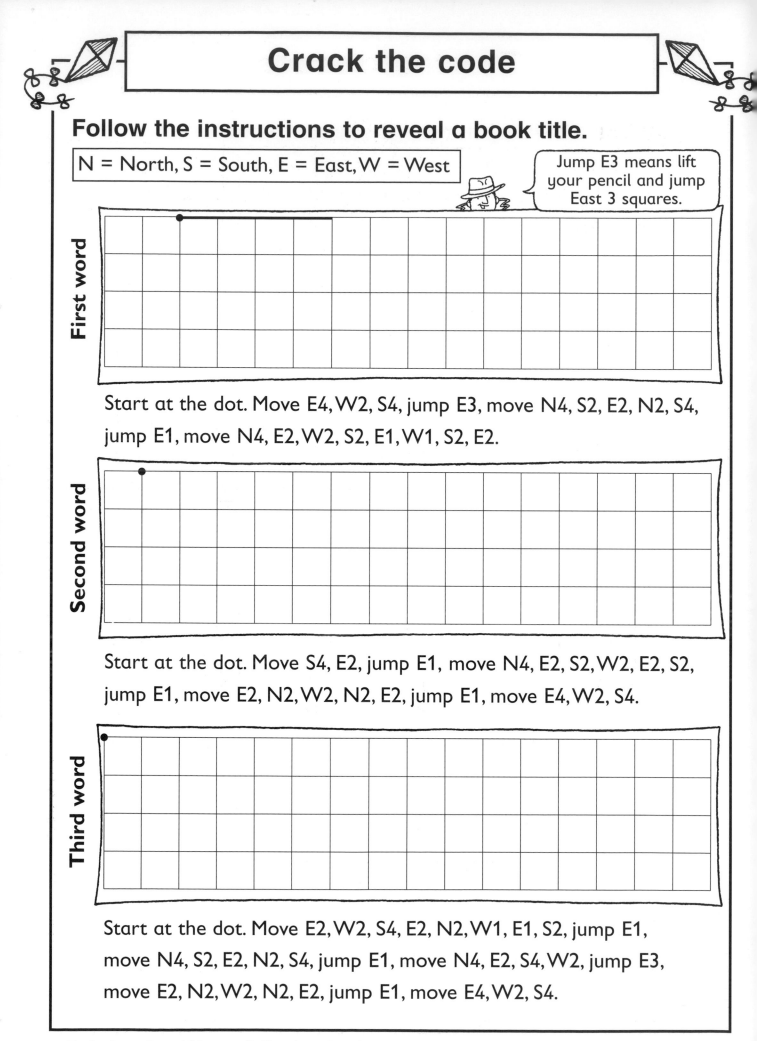

First word

Start at the dot. Move E4, W2, S4, jump E3, move N4, S2, E2, N2, S4, jump E1, move N4, E2, W2, S2, E1, W1, S2, E2.

Second word

Start at the dot. Move S4, E2, jump E1, move N4, E2, S2, W2, E2, S2, jump E1, move E2, N2, W2, N2, E2, jump E1, move E4, W2, S4.

Third word

Start at the dot. Move E2, W2, S4, E2, N2, W1, E1, S2, jump E1, move N4, S2, E2, N2, S4, jump E1, move N4, E2, S4, W2, jump E3, move E2, N2, W2, N2, E2, jump E1, move E4, W2, S4.

Teachers' note Some children may find it easier to draw the compass points onto the top of the sheet as a reference point. Encourage the children to use a coloured pencil and to cross off each instruction as they go to avoid confusion. As an extension activity, the children could be given squared paper and asked to write the code for a short title.

**Developing Numeracy
Measures, Shape and Space
Year 3**
© A & C Black 2001

54

Twist and score

• **Play this game with a partner.**

☆ Cut out the cards and the arrow at the bottom of the page.

☆ Place the arrow on the dot, pointing to 0 points. Spread the cards face down.

☆ Take turns to pick a card. Turn the arrow **clockwise** through the angle shown. Put the card back and mix them up.

☆ You score the number of points the arrow points to.

☆ The winner is the first to reach 30 points.

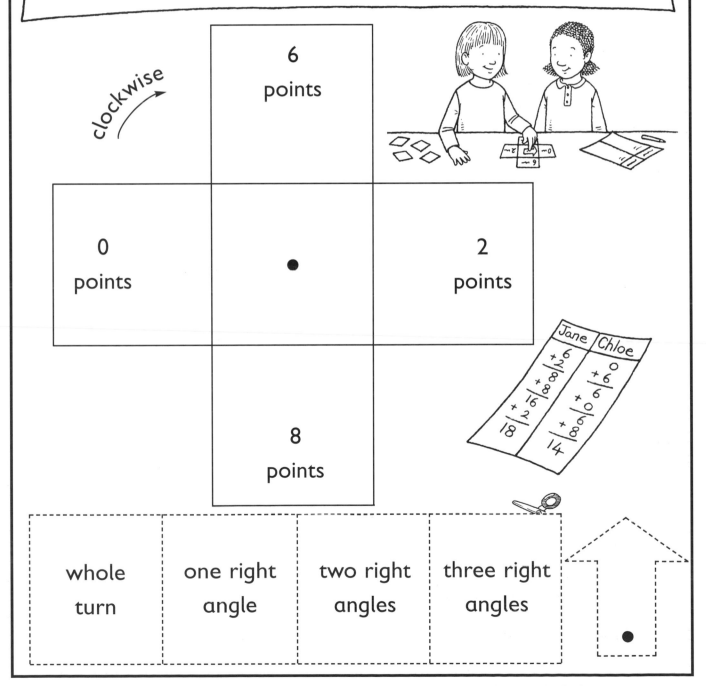

6 points

clockwise

0 points

2 points

8 points

Jane	Chloe
+6	0
+2	+6
+8	+6
16	+0
+2	+6
18	+8
	14

whole turn | one right angle | two right angles | three right angles

Teachers' note Each pair of children needs one copy of the sheet. The sheet could be copied onto card and laminated to provide a more permanent resource. The children could push a paper fastener through the dots to fasten the arrow to the game board.

Developing Numeracy
Measures, Shape and Space
Year 3
© A & C Black 2001

Turn tiles: 1

- **Cut out the tiles from the 'Tiles' sheet.**
- **Follow the instructions. Glue the tiles in the correct positions.**

> When the tile is in this position, it is called the **L tile**.

	Turn the L tile one right angle clockwise	Turn the L tile two right angles clockwise	
Turn the L tile one right angle clockwise	Turn the L tile one right angle anticlockwise	The L tile	Turn the L tile two right angles clockwise
The L tile	Turn the L tile two right angles anticlockwise	Turn the L tile one right angle clockwise	Turn the L tile three right angles clockwise
Turn the L tile three right angles anticlockwise	Turn the L tile three right angles clockwise	Turn the L tile one whole turn	Turn the L tile half a turn
Turn the L tile two whole turns	Turn the L tile half a turn	Turn the L tile one right angle clockwise	Turn the L tile one right angle anticlockwise
	Turn the L tile four right angles	Turn the L tile one right angle anticlockwise	

Teachers' note The children will need copies of page 58 for this activity. If necessary, demonstrate turning the tile clockwise, anticlockwise and through a number of right angles. As an extension, the children could draw lines of symmetry onto this pattern or make their own symmetrical patterns with the tiles and write turning instructions.

Developing Numeracy
Measures, Shape and Space
Year 3
© A & C Black 2001

Turn tiles: 2

- **Cut out the tiles from the 'Tiles' sheet.**
- **Follow the instructions. Glue the tiles in the correct positions.**

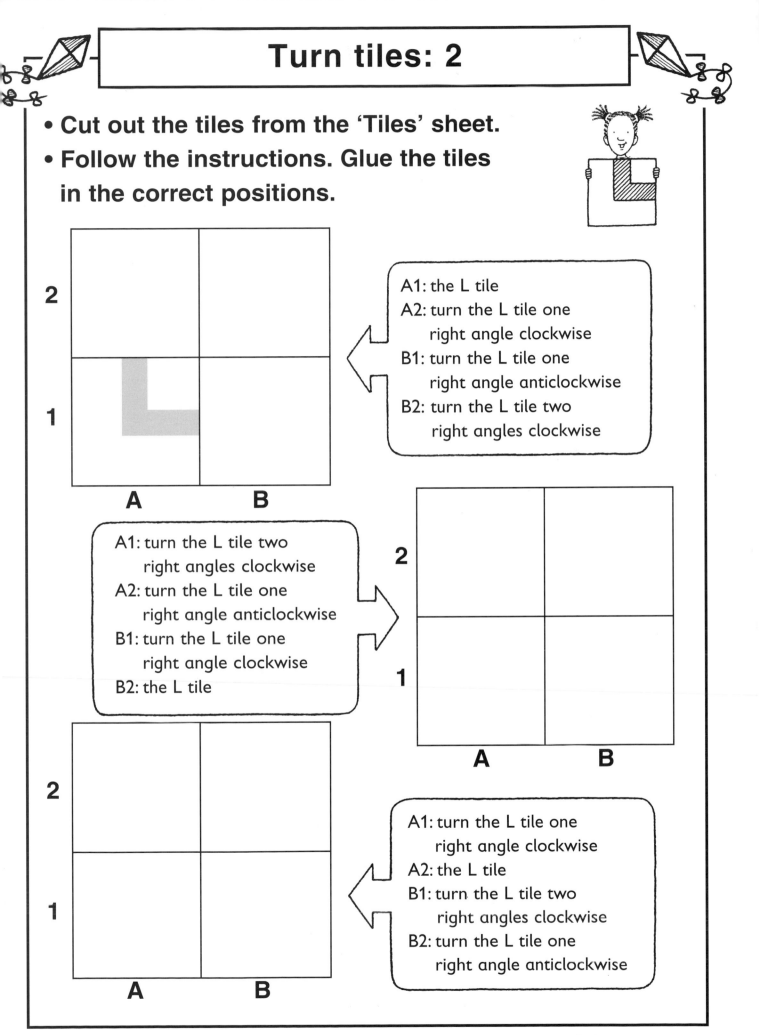

A1: the L tile
A2: turn the L tile one
   right angle clockwise
B1: turn the L tile one
   right angle anticlockwise
B2: turn the L tile two
   right angles clockwise

A1: turn the L tile two
   right angles clockwise
A2: turn the L tile one
   right angle anticlockwise
B1: turn the L tile one
   right angle clockwise
B2: the L tile

A1: turn the L tile one
   right angle clockwise
A2: the L tile
B1: turn the L tile two
   right angles clockwise
B2: turn the L tile one
   right angle anticlockwise

Teachers' note The children will need copies of page 58 for this activity. They should first complete the activity on page 56. As an extension, the children could make their own patterns with the tiles and write turning instructions. These could be used as part of a display on right angles. Similar activities can be carried out on a computer using more complex tiles.

**Developing Numeracy
Measures, Shape and Space
Year 3**
© A & C Black 2001

Tiles

Teachers' note This sheet should be used in conjunction with pages 56 and 57. One copy per child provides enough tiles for both activities. Point out that each tile is identical, but that it may look different when rotated.

Developing Numeracy
Measures, Shape and Space
Year 3
© A & C Black 2001

Routes

- **Follow this route on the map.**

Start at the park.

Go forward. Turn one right angle anticlockwise.

Go forward. Turn one right angle clockwise.

Go forward to the traffic lights.

Turn one right angle anticlockwise.

Go forward.

Finish at _____

Clockwise is a right turn and **anticlockwise** is a left turn.

left   right

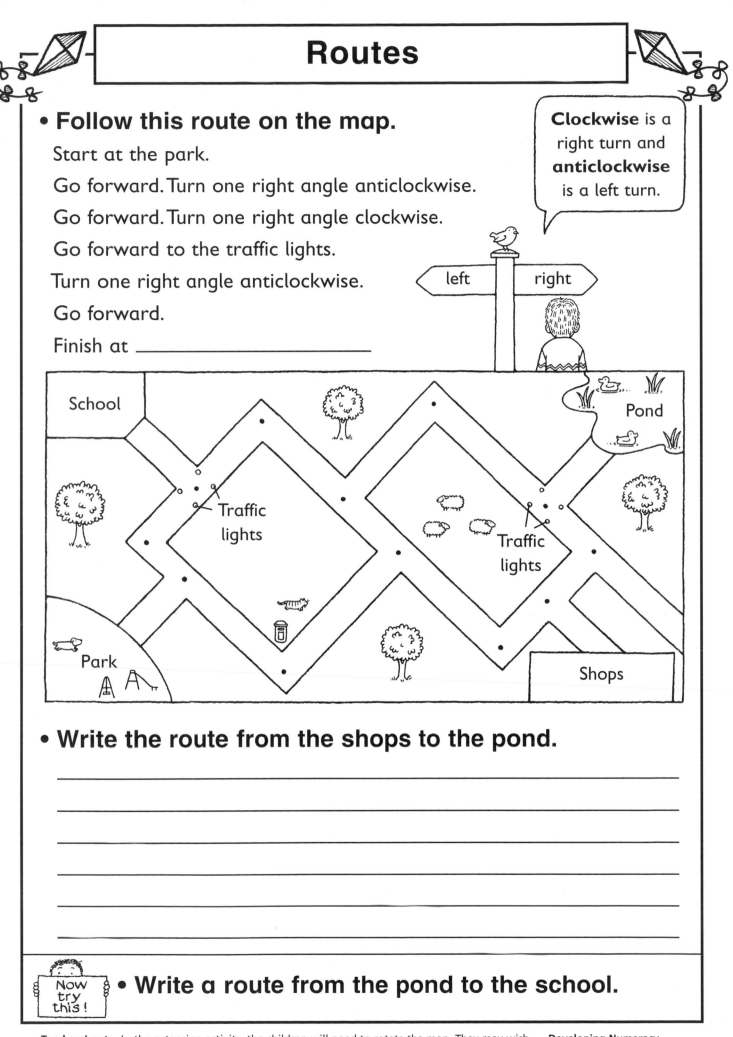

School

Pond

Traffic lights

Traffic lights

Park

Shops

- **Write the route from the shops to the pond.**

_____

_____

_____

_____

_____

Now try this!

- **Write a route from the pond to the school.**

Teachers' note In the extension activity, the children will need to rotate the map. They may wish to discuss ways of shortening the instructions. Introduce 90° if appropriate as a shorthand for 'right angle'.

Developing Numeracy
Measures, Shape and Space
Year 3
© A & C Black 2001

Animal angles

- **Trace this right angle onto tracing paper.**
- **Use it to test whether each angle is**
  larger **than,** smaller **than, or** exactly **a right angle.**

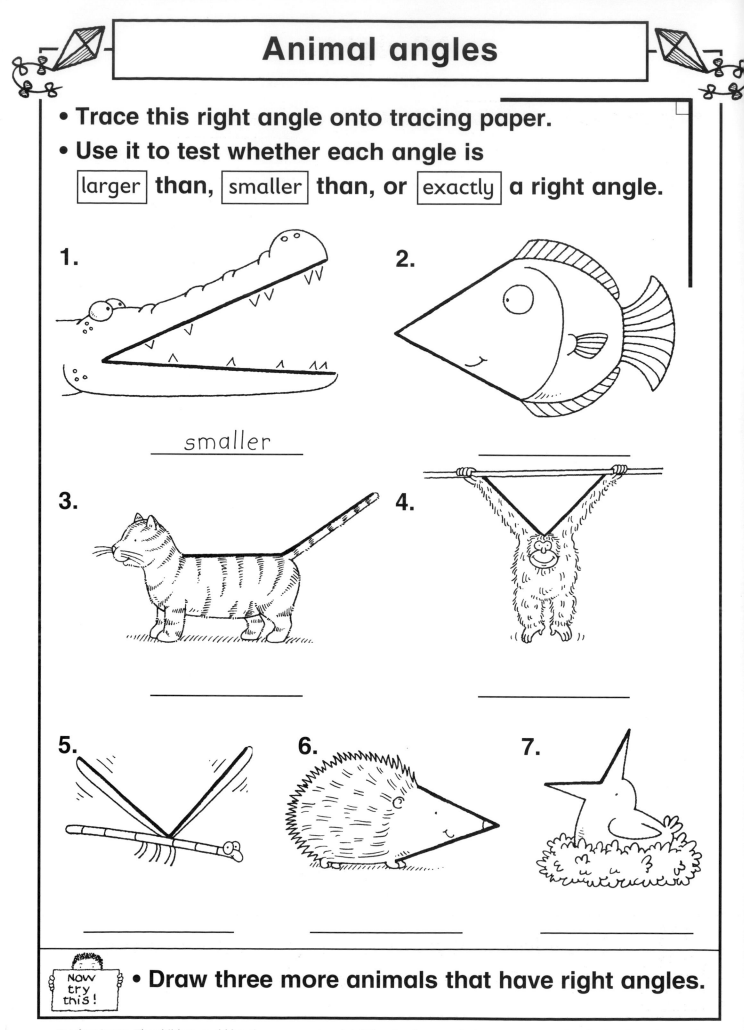

1.

2.

smaller

3.

4.

5.

6.

7.

NOW try this!
- **Draw three more animals that have right angles.**

Teachers' note The children could be given pre-prepared right angles drawn onto transparent paper for this activity. Demonstrate how to place the right-angle template over the animal angles to test their size, and explain that a right angle can be marked with a small square.

Developing Numeracy
Measures, Shape and Space
Year 3
© A & C Black 2001

All right?

- **Draw two more** flat **shapes in each set.**

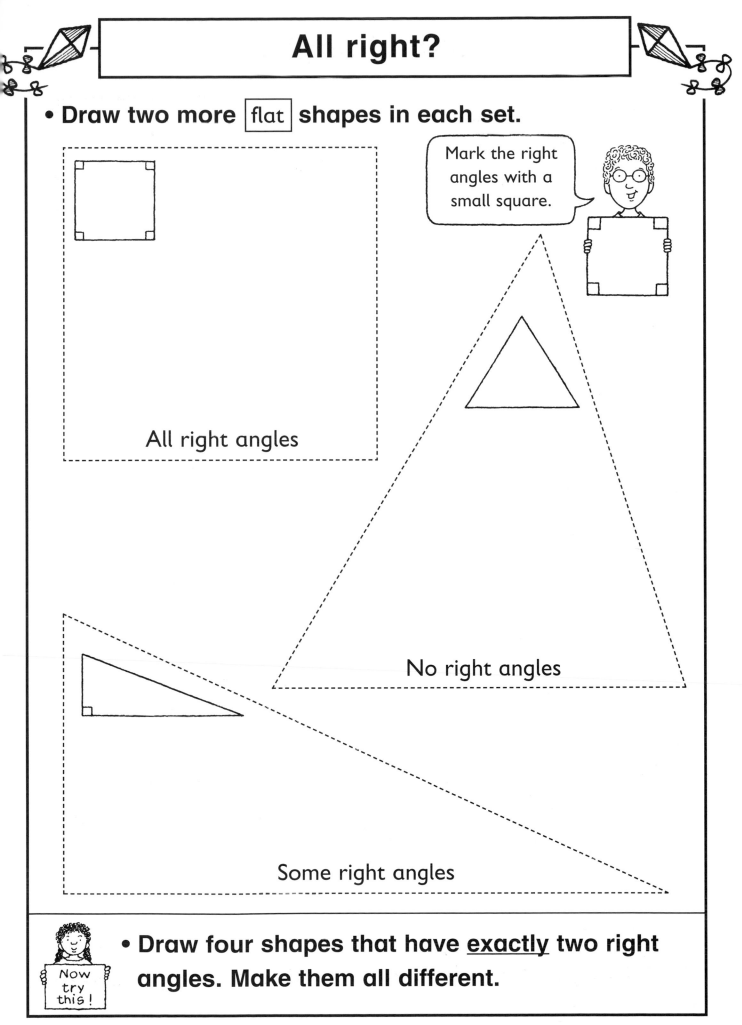

Mark the right angles with a small square.

All right angles

No right angles

Some right angles

- **Draw four shapes that have <u>exactly</u> two right angles. Make them all different.**

Now try this!

Teachers' note Some children may need a set of 2-D shapes to refer to, along with the right-angle template from page 60. Encourage the children to use a ruler when drawing the shapes.

**Developing Numeracy
Measures, Shape and Space
Year 3**
© A & C Black 2001

An angle tangle

- On each carriage, mark the smallest angle $\boxed{S}$ and the largest angle $\boxed{L}$.

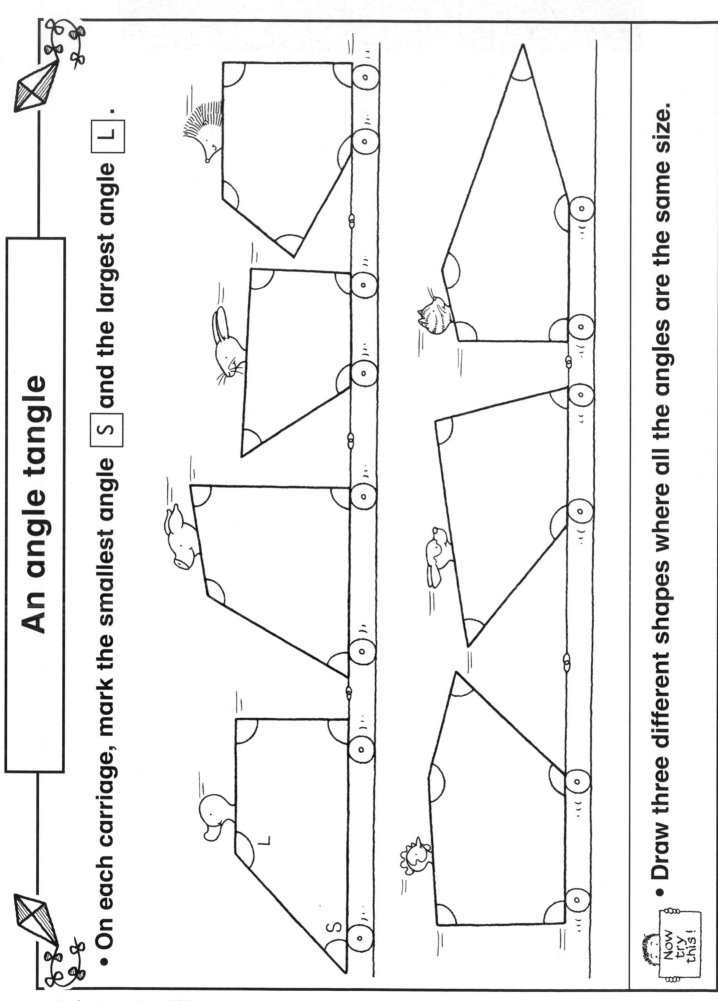

- Draw three different shapes where all the angles are the same size.

Teachers' note Some children struggle to visualise angles within shapes and to estimate their size. Provide the children with a right-angle template and encourage them first to predict whether each angle is smaller or larger than a right angle. A second copy of the sheet may help some children – by cutting out the carriages they can use them to check and compare the angles.

Developing Numeracy
Measures, Shape and Space
Year 3
© A & C Black 2001

Answers

p 7
1. kilometres
2. metres
3. centimetres
4. centimetres
5. metres
6. kilometres
7. centimetres
8. centimetres
9. metres
10. kilometres

p 8
1. 4 m
2. 15 cm
3. 12 cm
4. 1·5 m
5. 10 m
6. 50 cm
7. 15 cm

p 9
1. 4 m
2. 2 cm
3. 40 cm
4. 15 cm
5. 20 m
6. 2 m
7. 21 cm
8. 5 cm

p 10
Now try this!
2 m, 200 cm and 2·0 m
205 cm, 2·05 m
250 cm, 2·5 m

p 11
2·5 cm, 5 cm, 3·5 cm, 6·5 cm, 5·5 cm, 1·5 cm, 6 cm, 8·5 cm
Now try this!
3·5 cm

p 12
1. 14 cm
2. 17 cm
3. 19 cm
4. 14·5 cm
5. 15·5 cm

p 13
1. 20 cm
2. 18 cm
3. 19 cm
4. 23 cm
5. 25 cm

p 14
a 200 g
b 400 g
c 300 g
d 100 g
e 500 g
f 450 g
Now try this!
e, f, b, c, a, d

p 15
1. more than 1 kilogram
2. exactly 1 kilogram
3. less than 1 kilogram
4. less than 1 kilogram
5. exactly 1 kilogram
6. exactly 1 kilogram
7. exactly 1 kilogram
8. less than 1 kilogram

p 16
1. muffins
   muffins
   meringues
   muffins
   cookies
   straws
2. scones
   scones
   cookies
   meringues
   scones
   scones

p 17
Cheese scones
200 g	flour
50 g	butter
100 g	cheese
600 ml	milk

Cheese straws
100 g	butter
300 g	flour
4	eggs
200 g	cheese

Chocolate meringues
6	eggs
60 g	sugar
100 g	chocolate
150 g	cream

Crunchies
400 g	flour
400 g	butter
300 g	sugar

Choc-chip cookies
500 g	butter
300 g	sugar
600 g	flour
40 g	chocolate

Chocolate muffins
1 kg	flour
400 g	sugar
1 litre	milk
50 g	chocolate
2	eggs

Now try this!
Cheese scones
50 g	flour
12½ g	butter
25 g	cheese
150 ml	milk

Cheese straws
25 g	butter
75 g	flour
1	egg
50 g	cheese

Chocolate meringues
1½	eggs
15 g	sugar
25 g	chocolate
37½ g	cream

Crunchies
100 g	flour
100 g	butter
75 g	sugar

Choc-chip cookies
125 g	butter
75 g	sugar
150 g	flour
10 g	chocolate

Chocolate muffins
¼ kg	flour
100 g	sugar
250 ml	milk
12½ g	chocolate
½	egg

p 19
millilitres
millilitres
litres
litres
litres
litres
millilitres
millilitres or litres
millilitres or litres

p 20
1. 300 millilitres
2. 200 litres
3. 1 litre
4. 40 litres
5. 2 litres
6. 100 litres

p 22
metres	centimetres
grams	kilograms
millilitres	kilometres
litres	metres

p 23
Now try this!
120 minutes
48 hours
104 weeks
30 minutes
6 months

p 26
1. Friday
2. Thursday
3. 01/04/03
4. 06/04/03, 13/04/03, 20/04/03, 27/04/03
5. 31/03/03, Monday
6. 01/05/03, Thursday
Now try this!

M	T	W	T	F	S	S
			1	2	3	4
5	6	7	8	9	10	11
12	13	14	15	16	17	18
19	20	21	22	23	24	25
26	27	28	29	30	31	

p 27
Monday am, Tuesday am, Wednesday pm and pm, Thursday pm, Friday pm, Saturday pm, Sunday am, Monday am, Tuesday pm

p 28
1. five past six
2. twenty past seven
3. ten past eleven
4. twenty-five past nine
5. five to three
6. twenty to twelve
7. ten to seven
8. twenty-five to nine
9. twenty-five to two

Now try this!
1. twenty-five to seven
2. ten to eight
3. twenty to twelve
4. five to ten
5. twenty-five past three
6. ten past twelve
7. twenty past seven
8. five past nine
9. five past two

p 31
1. cylinder*
2. cone
3. triangular prism*
4. cuboid*
5. sphere
6. cube*
7. square-based pyramid
8. hemi-sphere
9. hexagonal prism*

* = prism

p 32
1. triangular prism
2. cylinder
3. cone
4. square-based pyramid
5. cuboid
6. hemi-sphere

p 34
Now try this!
The number of faces increases by one, i.e. one face, two faces, three faces, four faces...

p 45
2	1	2
1	2	0
2	0	1
2	1	2

p 46
0	1
0	2
1	0
1	1
2	2
1	

p 48
A1, C4, E2, E6
B5, C2, D3, F1
A6, B3, D5, F3

p 51
East 6 squares, North 1 square, West 1 square
West 3 squares, North 1 square, West 1 square
East 4 squares, North 1 square, West 3 squares, North 2 squares
South 1 square, East 4 squares (or vice versa)

p 52
South 2, East 2, North 1, East 3
South 3, West 3, South 1, West 1, North 4
South 3, East 4, South 1, West 1, North 2
(There are various alternative solutions.)

p 54
THE LAST GHOST

p 56
A continuous line should run around the grid.

p 57

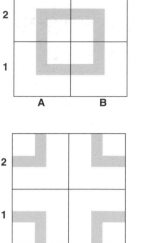

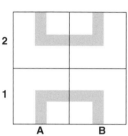

p 59
First route ends at the school.
Second route:
Go forward. Turn one right angle clockwise.
Go forward. Turn one right angle anti-clockwise.
Go forward to the traffic lights.
Turn one right angle clockwise.
Go forward to the pond.

Now try this!
Go forward. Turn one right angle clockwise.
Go forward. Turn one right angle anti-clockwise.
Go forward. Turn one right angle clockwise.
Go forward. Turn one right angle anti-clockwise.
Go forward to the traffic lights.
Turn one right angle clockwise.
Go forward to the school.

p 60
1. smaller
2. smaller
3. larger
4. exactly
5. exactly
6. smaller
7. larger

Learning Resource Centre Stockton Riverside College